Kinetics of Catalytic Reactions

M. Albert Vannice

Kinetics of Catalytic Reactions

With 48 Illustrations

 Springer

M. Albert Vannice
William H. Joyce Chaired Professor
Department of Chemical Engineering
The Pennsylvania State University
University Park, PA 16802
mavche@engr.psu.edu

ISBN-13: 978-1-4419-3758-2   e-ISBN: 978-0-387-25972-7   Printed on acid-free paper.

Printed in the United States of America.   (SPI/SBA)

9 8 7 6 5 4 3 2 1

springeronline.com

Foreword

Heterogeneous catalysis has shaped our past and will shape our future. Already involved in a trillion dollar's worth of gross domestic product, catalysis holds the key to near term impact areas such as improved chemical process efficiency, environmental remediation, development of new energy sources, and new materials. Furthermore, recent advances in understanding and computing chemical reactivity at the quantum level are opening new pathways that will accelerate the design of catalysts for specific functions. This enormous potential will ultimately be turned into reality in laboratory reactors and have its impact on society and the economy in the industrial reactors that lie at the heart of all chemical processes. Because the quantitative measure of catalyst performance is the reaction rate, its measurement is central to progress in catalysis.

The pages that follow are a comprehensive guide to success for reaction rate measurements and analysis in catalytic systems. The topics chosen, the clarity of presentation, and the liberal use of specific examples illuminate the full slate of issues that must be mastered to produce reliable kinetic results. The unique combination of characterization techniques, thorough discussion of how to test for and eliminate heat and mass transfer artifacts, evaluation of and validity tests for rate parameters, and justification of the uniform surface approximation, along with the more standard ideal reactor analyses and development of rate expressions from sequences of elementary steps, will enrich readers from both science and engineering backgrounds. Well-explained real examples and problems that use experimental data will help students and working professionals from diverse disciplines gain operational knowledge.

This book captures the career learning of an outstanding catalytic kineticist. Drawing on experience that began with a paper showing the power of physical chemical and thermodynamic constraints for eliminating incorrect rate formulations and includes a citation classic paper on the CO hydrogenation reaction over group VIII metals plus the development of rate expressions for a wide variety of catalyst systems, Vannice captures not only the theory of the Boudart school of chemical kinetics, but also its practical application. He has created a resource that will help the next generation of catalytic scientists and engineers provide the validated kinetic analyses that will be critical to the development of nano, micro, and macroscale catalytic systems of the future.

W. Nicholas Delgass
Purdue University
January 2005

Preface

The field of catalysis, especially heterogeneous catalysis, involves the utilization of knowledge from various disciplines, including chemical engineering, chemistry, physics, and materials science. After more than two decades of teaching a graduate course in catalysis, whose membership was comprised primarily of students from the above programs, and consulting with numerous industrial researchers, it became apparent to me that a book would be useful that focused on the proper acquisition, evaluation and reporting of rate data in addition to the derivation and verification of rate equations based on reaction models associated with both uniform and nonuniform surfaces. Such a book should familiarize and provide its reader with enough background information to feel comfortable in measuring and modeling heterogeneous catalytic reactions. For a single individual to attempt such an undertaking is almost a guarantee that some issues will be addressed less adequately than others; regardless, I hope that these latter topics will meet minimum standards! My goal is a text that will be self-contained and will provide a convenience for the practitioner in catalysis.

I would first like to acknowledge here the people who have been most influential in my life and have inspired me to this point in my career, whereupon I have been willing to undertake the effort to write this book (which is bound to reveal some of my deficiencies, I am sure!). Clearly, I must thank my parents who, during the time I attended a small high school of 19 students in Nebraska, always had me oriented towards a college education. Second, I gratefully acknowledge my graduate school mentor, Professor Michel Boudart, who created my interest in kinetics and catalysis and instilled in me the necessity of accurate, reproducible data. Third, I note my friends from graduate school, especially Professor Nicholas Delgass, who have continued to educate me during the past three decades. (They have upheld the old adage that one should never stop learning.) Fourth, I must mention my graduate students, who have provided me much pleasure, not only in their accomplishments as we worked together, but also in the successes they have subsequently achieved. In particular, I would like to express special appreciation to one of them, Dr. Paul E. Fanning, who graciously volunteered his time to very carefully proofread this textbook and offer valuable suggestions. Also, I would like to acknowledge the review of Chapter 6.3 and the comments offered by Dr. Evgeny Shustorovich, both of which were greatly appreciated. Next, I would like to thank my secretary, Kathy Peters, for her patience and persistence during the time she typed the drafts of these chapters as they traveled, at times uncertainly, via air mail between Alicante, Spain and Penn State. I must also express my gratitude to Professor Francisco Rodriguez-Reinoso for his hospitality at the University of Alicante during my one-year sabbatical stay there to write this book. Last, but certainly not least, I sincerely thank my wife, Bette Ann, for her patience and understanding during the many days and nights I was absent during the past three decades while working in the lab to establish a research program or attending necessary scientific meetings (invariably on her birthday!). She was also kind enough to offer her secretarial skills and proofread this book from a grammatical perspective.

Contents

xii   Contents

List of Symbols

1. Regular Symbols

Symbol	Meaning
A	preexponential factor
A	area
A, B, C	molecular species A, B, C ....
A_i	affinity for step i $(-\Delta G_i)$
a	pre-exponential term in constant C in BET eq. (Chap. 2), area per volume (Chap. 4)
a_i	thermodynamic activity of species i (Chap. 6)
a, b, c	stoichiometric coefficients (Chap. 2)
a_i, b_i, c_i	reaction orders for species A, B, C, ...
a,b	constants in various equations
A_o, B_o	constants in Temkin isotherm (Chap 5)
b	constant in L-J expression (Chap. 5)
C	constant in BET equation
C_i, [i]	concentration of species i
C_o	bulk concentration
c	constant in Freundlich isotherm (Chap. 5)
D	molecular diffusivity
D	bond energy
D_{AB}	dissociation energy of molecule AB
D_M	metal dispersion
Da	Damköhler number
d, $\bar{d}$	diameter, average particle diameter
E	activation energy
E_{pot}	potential energy
F_i	molal flow rate of species i
f	fractional conversion
f	dimensionless concentration (Chap. 4)
f	width of surface nonuniformity, $t_o - t_i$ (Chap. 8)

G	Gibbs free energy
ΔG	Gibbs free energy change for reaction or adsorption
g_i	degeneracy of energy level
H	enthalpy
ΔH	enthalpy change for reaction or adsorption
h	heat transfer coefficient (Chap. 4)
h	Planck constant
K	constant in Scherrer equation (Chap. 3)
K	equilibrium constant
K_i	adsorption equilibrium constant for species i
k	rate constant
k_B	Boltzmann constant
k_g	mass transfer coefficient
L	active site density
L_i	enthalpy of vaporization of component i
ℓ	liter
M	metal atom
M_i	molecular weight of species i
m	mass
m	the value of $\alpha - \gamma$ in Temkin's eq. (Chap. 8)
N	TOF (turnover frequency)
N	number of atoms in a molecule (Chap. 6)
N_i	molar quantity of species i
N_{W-P}	Weisz-Prater number
n	number of catalytic turnovers (Chap. 2)
n	reaction order
n	coordination number (Chap. 6)
n_i	fraction or number of type i sites (Chap. 5,8)
n_s	site density (Chap. 3)
n	adsorbate uptake
n_m	monolayer adsorbate uptake in BET eq., maximum adsorbate uptake
P	pressure
P_c	critical pressure
P_o	saturation vapor pressure
Q	total partition function
Q	bond energy (Chap. 6)
Q_p, Q_c, Q_{ad}	Heats of physisorption, chemisorption and adsorption, respectively
q_i	partition function for i^{th} type of energy
R	gas constant
R_p	particle radius
$\Re$	global rate of reaction
r,r	rate of reaction
r	radius, distance (Chap. 6)

Re	Reynolds number
S	number of sites
S	selectivity
S	total surface area (Chap. 5)
S	entropy
ΔS	entropy change for reaction or adsorption
S or *	active site
s	second
s	sticking probability (Chap. 5)
s	surface domains
SV	space velocity
T	temperature
T_c	critical temperature
t	time
t	dimensionless affinity (A°/RT) (Chap. 8)
u	ratio of empty to filled sites (Chap. 8)
V	volume
V_i	molar volume of component i
V_m	monolayer volume adsorbed, BET equation
V_o	volumetric flow rate
v	velocity
$\bar{v}$	void volume (Chap. 4)
W	weight
x	mole fraction
x	distance (Chap. 4)
x	bond order (Chap. 6)
y	distance
Z	site coordination number (Chap. 5)

2.   Greek Symbols

Symbol	Meaning
α	transfer coefficient in Brønsted relation, constant for a Temkin surface, (Chap. 8)
α_{12}	thermodynamic correction term (Chap. 4)
β	line broadening of XRD peak (Chap. 3)
β	maximum decrease in concentration gradient in pores (Chap. 4)
γ	E_t/RT_s (Chap. 4)
γ	constant in Freundlich isotherm
γ_i	activity coefficient of component i
δ	monolayer thickness (Chap. 3)

δ	film thickness
ε	catalyst porosity
ε_i	energy of the ith state
$\eta, \bar{\eta}$	effectiveness factors
η_i	viscosity of component i
η^n	molecular coordination on a surface (n = # of bonds) (Chap. 6)
θ	contact angle (Chap. 3)
θ	fractional surface coverage
θ	diffraction peak angle
λ	radiation wave length (Chap. 3)
λ	thermal conductivity (Chap. 4)
λ	mean free path
μ_n	metal coordination sphere (n = # metal atoms) (Chap. 6)
ν_i	stoichiometric coefficient for species i
ν	vibrational frequency
ξ	extent of reaction
ρ	density
ρ	dimensionless radius (Chap. 4)
ρ_k	constant in Hammett relation (Chap. 8)
σ	surface tension
σ	steric factor (Chap. 5)
σ	molecular radius
σ	value for substituent group in Hammett relation (Chap. 8)
τ	space time
τ	catalyst tortuosity
Φ	Thiele modulus
χ	solvent association parameter

3. Subscripts

Symbol	Meaning
a	area
ad	adsorbed, adsorption
ap	apparent, applied (Chap. 3)
B	Boltzmann
b	bulk
c	catalyst, chemisorption
col	collision
D	diffusion
des	desorption
eff	effective
F	fractional

f	fluid
g	gas
i	species i
Kn	Knudsen
L,l	liquid
M	metal atoms, molecules (Chap. 5)
m	mass, Michaelis (Chap. 9)
m	mixture, maximum
n	number
ob	observed
p	pore, particle, physisorption
R	relative
r	reactor
RDS	rate determining step
s	surface
t	total, true (Chap. 4)
v	volume, vacant
W-P	Weisz-Prater
0 or o	initial (or inlet) value, bulk

1
Introduction

Catalysis is the phenomenon in which a relatively small amount of a foreign material, called a catalyst, increases the rate of a chemical reaction without itself being consumed. Although widely utilized now in many industrial processes, catalysis was not even recognized until the 19th century when Berzelius introduced the term in 1836. Other early pioneers in this field during this century include Davy, Faraday, Bertholet, Ostwald, and Sabatier [1–3]. Tremendous advances in catalytic processing were made at the beginning of the 20th century as hydrogenation of oils, fats, and waxes to food stuffs, ammonia synthesis from N_2 and H_2, ammonia oxidation to nitric acid, and the synthesis of hydrocarbons and organics from H_2 and CO were developed. However, the modeling of quantitative kinetics of reactions occurring on catalytic surfaces essentially did not begin until the contributions of Langmuir between 1915 and 1920, which provided a relationship between adsorbed species and measurable experimental parameters [2,4]. This approach was subsequently broadened and utilized by Hinshelwood, Taylor, Rideal, Eley, Hougen, Watson and others who were among the first to determine and model the kinetics of reactions on heterogeneous catalysts. This book reviews, analyzes and builds on these earlier models.

Catalysis is an interdisciplinary field, and a thorough study of catalytic reactions requires knowledge from chemistry, physics, mathematics, chemical engineering and materials science, for example. Few, if any, investigators can claim to be an "expert" in all these fields, but such a requirement is not necessary; however, one must have sufficient understanding of each discipline to: a) properly design kinetic experiments, b) satisfactorily characterize catalysts, c) acquire valid rate data and correctly express it, d) test for artifacts in these kinetic data, e) propose reasonable reaction models, f) derive proper rate expressions based on these models and, finally, g) evaluate the physical and thermodynamic consistency of the fitting parameters contained in these rate equations. The background information required for all these capabilities is very diverse and it is infrequently contained in detail in a single book. The goal here is to provide between these covers sufficient information about the techniques required, the theory behind them, and their appropriate utilization, not only to allow a practitioner in heterogeneous catalysis to properly conduct kinetic studies of catalyzed reactions, but also to provide an overview of the kinetics of catalyzed reactions for a graduate course, presumably in chemical engineering or chemistry.

The chapters in this book are based on notes used in a graduate course in heterogeneous catalysis that the author taught at Penn State, with continuous updating, for over 25 years [5]. They contain numerous examples and illustrations of catalyst characterization, reaction modeling, and rate law evaluation largely derived from the research undertaken in his laboratory during this period of time. One advantage of such an approach is that the chemisorption techniques described for the characterization of catalysts, dispersed metal catalysts in particular, include some of the most recent methods reported in the open literature. Another reason for the inclusion of a relatively large number of examples for kinetic modeling of real systems, both as illustrations and as homework problems, is the scarcity of such problems in most textbooks on kinetics and/or reactor design. If nothing else, this book will provide examples of consistent kinetic models and rate expressions that have successfully fit experimental rate data acquired from a variety of catalyzed reactions.

The approach taken in this book is largely a continuation of the "Boudart school of kinetics". The academic geneology of this school is quite interesting and worthy of mention. Michel Boudart's advising professor at Princeton University was Sir Hugh Taylor. Taylor, in turn, worked with Basset in Liverpool, Arrhenius in Stockholm, and Bodenstein in Hannover to obtain his Ph.D. Arrhenius worked with Ostwald, Kohlrausch, Boltzmann and van't Hoff after receiving his doctorate, while Bodenstein also conducted postdoctoral research with Ostwald [2]. Clearly, a solid foundation in thermodynamics, kinetics and catalysis was established by this research lineage.

If I were to take the liberty to summarize Boudart's philosophy about kinetics and catalysis, I would do so as follows:

A) First, obtain reproducible experimental rate data.
B) These data must then be tested and checked to verify the absence of artifacts, such as mass and heat transfer limitations.
C) A catalytic cycle, preferably comprised of a series of elementary steps, is proposed.
D) Assumptions are made regarding dominant surface species and relative rates of the elementary steps to allow the derivation of a rate expression consistent with the data. This process also includes the choice of an ideal or nonideal catalytic surface.
E) The kinetic and adsorption equilibrium constants contained in the rate expression are evaluated using a set of guidelines to verify they are physically reasonable and thermodynamically consistent.

F) Based on this reaction model, additional tests of its validity can be proposed and conducted, if possible.

G) The catalyst is characterized to determine the active surface area and, if possible, to count the number of active sites. Additional information about the chemical state of the working catalyst is also very desirable, if it can be obtained.

Having a rate expression that is consistent with data taken in the kinetic control regime and which contains physically and thermodynamically consistent parameters still does not guarantee that the reaction model upon which it is based is the correct one. However, it does show that the model *could* be the correct one, and it provides much more information than a simple power rate law because the model gives insight into the state of the working catalyst. This provides some knowledge about the catalytic processes on the surface and allows reasonable extrapolation of the model to conditions outside of the experimental region examined.

Use of the information in these chapters will allow a researcher conducting experiments with catalysts in either an industrial or an academic laboratory to assess their results and determine the presence or absence of heat and mass transfer effects. Proper catalyst characterization provides the capability to report kinetic results properly in the form of specific or normalized activity, preferably in the form of a turnover frequency. The utilization and justification of reaction models based on uniform or ideal surfaces is discussed in detail, and numerous examples are provided. However, kinetic rate expressions based on the premise of nonuniform surfaces are also examined in depth to provide an alternate route to obtain a rate law, should the investigator wish to do so. In most studies of catalyzed reactions, the kinetics of these reactions lie at the heart of the investigation, not only because accurate comparisons of performance among different catalysts must be obtained, but also because accurate rate expressions can provide insight about the surface chemistry involved and they must be available for proper reactor design.

It is worthwhile to mention here several topics that are NOT going to be discussed in any detail. This book is oriented toward the typical investigator in catalysis, who has access to readily available experimental tools and techniques; thus, catalyst characterization based on ultra high vacuum (UHV) techniques, Mössbauer spectroscopy, electron paramagnetic resonance (EPR) spectroscopy, nuclear magnetic resonance (NMR) spectroscopy and magnetic susceptibility is discussed little or not at all, and extended x-ray absorption fine structure (EXAFS) is mentioned but not discussed in detail. This is because only a small fraction of researchers have ready access to these methods. Also, the microkinetic approach to rate expressions [6] is not discussed, even though the resulting rate laws are admittedly preferred when they can be accurately obtained. This choice was made not only because this approach depends so heavily on rate constants obtained using

UHV systems, which are seldom available for such measurements and are time and cost intensive, but also because these more detailed microkinetic rate laws typically simplify to more conventional rate expressions over a chosen range of reaction conditions [6].

One final comment should be made to facilitate reading this book. Chemical reactions are identified by numbers on the left margin whereas equations are identified by numbers on the right margin. Also, these numbers in illustrations and problems are distinguished by being italicized and do not include the chapter number.

References

1. R. L. Burwell, Jr., *Chemtech*, 17 (1987) 586.
2. K. J. Laidler, "Chemical Kinetics", 3rd ed., Harper & Row, NY, 1987.
3. J. M. Thomas and W. J. Thomas, "Principles and Practice of Heterogeneous Catalysis", VCH, Weinheim, 1997.
4. I. Langmuir, *J. Am. Chem. Soc.*, 40 (1918) 1361.
5. M. A. Vannice, *Chem. Eng. Education*, Fall, 1979.
6. J. A. Dumesic, D. F. Rudd, L. M. Aparicio, J. E. Rekoske and A. A. Trevino, "The Microkinetics of Heterogeneous Catalysis", Am. Chem. Soc., Washington, D.C., 1993.

2
Definitions and Concepts

It is important that precise and unambiguous terms be used when dealing with rates of reaction and reaction modeling of a chemical system. Many of the definitions provided here have been taken from those provided by the IUPAC [1].

2.1 Stoichiometric Coefficients

A balanced chemical reaction can be expressed as

(2.1) $$0 = \Sigma_i \nu_i B_i$$

where ν_i is the stoichiometric coefficient (positive for products and negative for reactants) of any product or reactant B_i. Thus an example reaction between A and B

(2.2) $$aA + bB \longrightarrow cC + dD$$

can be expressed as

(2.3) $$0 = -aA - bB + cC + dD$$

2.2 Extent of Reaction

This quantity is defined as

$$\xi(mol) = (N_i - N_{i_o})/\nu_i \qquad (2.1)$$

where N_{i_o} and N_i are the quantities of substance B_i, expressed in mole, at time zero and at any other time, respectively; consequently, ξ can be viewed as a mole of reaction.

2.3   Rate of Reaction

An unambiguous rate of reaction, r, is defined by the number of occurrences
of this stoichiometric event, such as that shown by reaction 2.2, per unit
time. For a particular species, i, its rate of production, r_i, is related to r by the
stoichiometric coefficient, i.e.,

$$r_i = v_i r \tag{2.2}$$

Rates of reaction can be expressed in terms of process variables associated
with a given reactor type via relationships generated by material balances on
that reactor. Because rate measurements are essentially always made in a
reactor, a discussion of the rate of reaction can be initiated by considering a
well-mixed, closed reactor system typically referred to as a batch reactor. In
this system, the advancement of the reaction is measured by the molar extent
of reaction, ξ, and the reaction rate is equivalent to the rate of change of the
molar extent of reaction, i.e.,

$$r = \dot{\xi} = d\xi/dt = v^{-1}dN_i/dt \ (mol/time) \tag{2.3}$$

where $r_i - dN_i/dt$ is the rate of formation (or disappearance) of compound
B_i. To make this an intrinsic property, it is normalized to unit reactor volume
to get r (mol time^{-1} volume^{-1}) or, when a catalyst is used, to the unit volume,
or unit mass, or unit area of the catalyst. Thus, choosing seconds as the unit of
time, one can define a volumetric rate:

$$r = r_v = \left(\frac{1}{V}\right)d\xi/dt \ (mol \ s^{-1} \ cm^{-3}) \tag{2.4}$$

where V should be only the volume of the catalyst particles excluding the
interparticle volume; or a specific rate:

$$r_m = \left(\frac{1}{m}\right)d\xi/dt \ (mol \ s^{-1} \ g^{-1}) \tag{2.5}$$

where m is the mass of the catalyst; or an areal rate

$$r_a = \left(\frac{1}{A}\right)d\xi/dt \ (mol \ s^{-1} \ cm^{-2}) \tag{2.6}$$

where A is the area of the catalyst. It should be specified whether A is the
total surface area or the surface area of only the active component, such as
the metal surface area of a dispersed metal catalyst. All subsequent rates in
this book will be some form of an intrinsic rate.

2.4   Turnover Frequency or Specific Activity

When the reaction rate is normalized to the surface area of the active
component in the catalyst, such as the metal surface area as just mentioned,
it is frequently referred to as the specific activity. If this areal rate is further

normalized to the number of surface metal atoms present, or to another specified type of site that has been counted by some stated method, then a turnover frequency (TOF), based usually on a specified reactant, is obtained

$$\text{TOF} = \frac{1}{S} = \frac{N_{Av}}{S} dN_i/dt \qquad (2.7)$$

where N_{Av} is Avogadro's number $(6.023 \times 10^{23}$ molecules/g mole) and S represents the number of sites in the experimental system and can be represented as

$$S = LA \qquad (2.8)$$

where L is the number density of sites (per unit area, such as cm^{-2}). A TOF has units of reciprocal time and is typically expressed as s^{-1}.

Several aspects must be emphasized at this time. First, for all these representations of rate, all conditions of temperature, initial concentrations or partial pressures, and extent of reaction must be specified. Second, for appropriate comparisons among different catalysts, areal rates or TOF values must be reported to correct for variations in active surface area. Finally, precise TOFs for heterogeneous catalysts are not so readily definable as those in homogeneous or enzyme catalysis because adsorption sites typically measured by the chemisorption of an appropriate gas and used to count surface metal atoms, for example, do not necessarily correspond to 'active' sites under reaction conditions on a one-to-one basis. The exact atom or grouping of atoms (ensemble) constituting the active site is typically not known for any heterogeneous reaction and, in fact, it is very likely that a variety of active sites may exist, each with its own rate, thus the observed TOF then represents an average value of the overall catalyst activity. Regardless, if rates are normalized to the number of surface metal atoms, M_s, in a metal catalyst as determined by some adsorption stoichiometry, for example, this not only provides a lower limit for the true TOF, but it also allows meaningful comparison among various catalysts, as stated above, as well as rate data obtained in different laboratories. *Thus TOFs (or areal rates) must be reported whenever possible in any proper catalytic study.*

The number of times, n, that the overall reaction takes place through the catalytic cycle represents the number of turnovers, and the rate is then [2]

$$r = dn/dt(s^{-1}) = N_{Av}d\xi/dt \qquad (2.9)$$

The turnover frequency, N, based on this value and expressed as s^{-1}, is:

$$N(s^{-1}) = \frac{1}{S} dn/dt \qquad (2.10)$$

and its relationship to that defined by equation 2.7 is therefore

$$N = 1/\nu_i \, dN_i/dt \qquad (2.11)$$

The number of turnovers a catalyst can produce is the best way to define the life of a catalyst, and in real systems this number can be very large, frequently exceeding 10^6 [2]. Because TOFs used in industrial processes are frequently near $1\,s^{-1}$, lifetimes of one month to one year can readily be achieved.

2.5   Selectivity

The term selectivity, S, is used to describe the relative rates of two or more competing reactions on a catalyst. Such competition includes different reactants undergoing simultaneous reactions or a single reactant involved in two or more reactions. In the latter case, a fractional selectivity, S_F, for each product is defined by the equation

$$S_F = \dot{\xi}_i/\Sigma_i\dot{\xi}_i = r_i/\Sigma_i r_i \qquad (2.12)$$

and a relative selectivity, S_R, for each pair of products is defined by

$$S_R = \dot{\xi}_i/\dot{\xi}_j = r_i/r_j \qquad (2.13)$$

2.6   Structure-Sensitive and Structure-Insensitive Reactions

For some reactions on metal surfaces, the activity of the catalyst depends only on the total number of surface metal atoms, M_s, available, and these are termed structure-insensitive reactions; consequently, the TOF is essentially independent of metal dispersion or crystal plane and varies over a very small range (within a factor of 5, for example). For other reactions, the TOF is much greater on certain surface sites, thus the activity can be dependent on metal dispersion, crystal plane, or defect structures. These are termed structure-sensitive reactions. Preceding these terms, such reactions have been referred to as facile and demanding reactions, respectively [3].

2.7   Elementary Step and Rate Determining Step (RDS)

An elementary step is a reaction written exactly as it occurs at the molecular level, and the stoichiometry defined by reaction 2.1 does not describe how the chemical transformation occurs unless it represents an elementary step. Thus an arbitrary choice of stoichiometric coefficients cannot be made; for example, the dissociative adsorption of a diatomic molecule such as H_2 must be written

$$(2.4) \qquad\qquad H_{2(g)} + 2* \rightleftharpoons 2H* $$

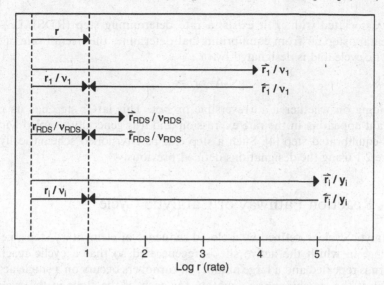

FIGURE 2.1. Relative rates and the net rate in a catalytic cycle, where RDS indicates a rate determining step.

and it cannot be represented in a kinetic sequence as

$$1/2H_{2(g)} + * \rightleftharpoons H* \tag{2.5}$$

in contrast to equations representing the equilibrium thermodynamics of reactions. Here * represents an active site involved in the catalytic sequence describing the overall reaction. The net rate of an elementary step, r, is the difference between the forward rate, $\vec{r}$, and the reverse rate, $\overleftarrow{r}$:

$$r = \vec{r} - \overleftarrow{r} \tag{2.14}$$

This step can be reversible when $\vec{r} \simeq \overleftarrow{r}$, which is represented by a double arrow:

$$\text{Reactants} \rightleftharpoons \text{Products}$$

or it can be irreversible if $\vec{r} \gg \overleftarrow{r}$, and this is represented by a single arrow:

$$\text{Reactants} \longrightarrow \text{Products}$$

Finally, this step may be essentially at equilibrium if $\vec{r}$ and $\overleftarrow{r}$ are both very large compared to the slow step(s) and if $\vec{r} \cong \overleftarrow{r}$, thus this quasi-equilibrated step is denoted by [2]:

$$\text{Reactants} \rightleftharpoons\!\!\!\ominus\!\!\!\rightleftharpoons \text{Products}$$

Note that under reaction conditions with a net forward rate, this step cannot be exactly at equilibrium; it only requires that both the forward and reverse rates occur much more rapidly, typically orders of magnitude greater, than those associated with, if it exists, a rate determining step (RDS), i.e., an elementary step far from equilibrium that determines the overall rate of the catalytic cycle and is designated by

$$-A\!\!\rightarrow \text{ or } \cancel{A}$$

depending on whether it's irreversible or not. This latter step has its rate constant appearing in the rate expression and it is generally coupled with a quasi-equilibrated step [4]. Such a step can be envisioned schematically in Figure 2.1 using the designations defined previously.

2.8   Reaction Pathway or Catalytic Cycle

A catalytic cycle is defined by a closed sequence of elementary steps, i.e., a sequence in which the active site is regenerated so that a cyclic reaction pattern is repeated and a large number of turnovers occurs on a single active site [5]. If the stoichiometric equation for each of the steps in the cycle is multiplied by its stoichiometry number, i.e., the number of times it occurs in the catalytic cycle, and this sequence of steps, the reaction pathway, is then added, the stoichiometric equation for the overall reaction is obtained. This equation must contain only reactants and products because all intermediate species must cancel out, and this overall reaction is represented by an equal sign:

$$\text{Reactants} = \text{Products}$$

with $\rightleftharpoons$ used if it is equilibrated and $\Rightarrow$ used if it is far from equilibrium [2].

It should be stressed that a catalytic sequence representing the reaction pathway may not contain a RDS. As an example, consider the catalytic gas-phase oxidation of carbon monoxide. On some metals the surface reaction between adsorbed CO molecules and O atoms is the RDS, then, if * is an active site, the sequence can be represented as:

$$2[CO + * \overset{K_{CO}}{\rightleftharpoons} CO*]$$

$$O_2 + 2 * \overset{K_{O_2}}{\rightleftharpoons} 2O*$$

$$2[CO* + O* \overset{k}{-\!\!A\!\!\rightarrow} CO_2* + *]$$

$$2[CO_2* \overset{K'}{\rightleftharpoons} CO_2 + *]$$

$$\overline{}$$

$$2CO + O_2 \implies 2CO_2$$

Here 2 represents the appropriate stoichiometric number in the cycle and all the reactive intermediates — *, CO* and O* — cancel out to give the bottom overall equation. Alternatively, on other metals a RDS does not exist under certain conditions and the above sequence can become:

$$2[CO + * \underset{K_{CO_2}}{\rightleftharpoons} CO*]$$

$$O_2 + 2 * \xrightarrow{k_1} 2O*$$

$$2[CO* + O* \xrightarrow{k_2} CO_2 + 2*]$$

$$\overline{2CO + O_2 \implies 2CO_2}$$

2.9   Most Abundant Reaction Intermediate (MARI)

If, under reaction conditions, one of the adsorbed species dominates on the surface and the fractional coverage of this intermediate on the catalytic sites is much greater than any other species, then it is said to be the most abundant reaction intermediate (MARI). Technically, it may not be the most abundant surface intermediate (MASI) because some adsorbed species may not be participating in the reaction sequence [2], although these two terms tend to be used interchangeably [1].

2.10   Chain Reactions

A chain reaction is a closed sequence which is created by the formation of active centers due to the thermal decomposition of a molecular species or to some external source such as light or ionizing radiation. A chain reaction must consist of at least four steps: one for initiation, one for termination and at least two for chain propagation, with the last steps being the principal pathway for product generation.

2.11   Reaction Rates in Reactors

In flow reactors, various quantities are related to the reaction rate. One important one is the space velocity, which is defined by the volumetric flow rate of the reactant stream, V_o, specified at the inlet conditions of temperature and pressure with zero conversion (unless otherwise noted), and the catalyst volume, V_c, to be:

$$\text{Space velocity (SV)} = V_o/V_c(\text{time}^{-1}) \tag{2.15}$$

In designing reactors, the reactor volume, V_r, which is required to hold a given mass or volume of catalyst, is routinely used:

$$SV = V_o/V_r (\text{time}^{-1}) \tag{2.16}$$

thus V_r depends on the packing density of the catalyst particles. This quantity of SV is typically expressed in reciprocal hours, h^{-1}, and is frequently near unity in commercial processes. The inverse of the space velocity is the space time, τ

$$\tau = 1/SV = V_r/V_o \text{ (time)} \tag{2.17}$$

and it gives the time required to process one reactor volume of feed. The space time yield refers to the quantity of product produced per quantity of catalyst per unit time. It should be emphasized that the space time, τ, is equal to the average residence time, $\bar{t}$, only if all the following conditions are met: 1) P and T are constant throughout the reactor, 2) the density of the reacting mixture is independent of ξ, and 3) V_o is the reference volumetric flow rate [6].

2.12 Metal Dispersion (Fraction Exposed)

The dispersion, D_M, or fraction exposed of a metal catalyst is the ratio of the number of surface metal atoms to the total number of metal atoms:

$$D_M = N_{M_s}/N_{M_t} \tag{2.18}$$

i.e., the fraction of metal atoms at the surface, where N_{M_s} and N_{M_t} are typically reported per g catalyst.

2.13 Metal-Support Interactions (MSI)

A variety of metal-support effects can occur to alter the adsorptive and/or catalytic behavior of a metal surface, and these include: 1) Incomplete reduction of the metal; 2) Support-induced cluster size; 3) Epitaxial growth; 4) Particle morphology; 5) Contamination by the support; 6) Bifunctional catalysis; 7) Spillover and porthole phenomena; and 8) Charge transfer between a metal and a semiconductor [2]. In addition, one might cite the stabilization of extremely small (1-3 atom) metal clusters on a support [7].

Also, there is one additional type of metal-support effect that was originally termed SMSI (Strong Metal-Support Interactions) by the researchers at Exxon, where it was discovered [8], and its presence using a reducible oxide support was demonstrated by a marked decrease in H_2 and CO chemisorption capacity, especially the former, with no increase in metal crystallite size, i.e., no decrease in dispersion [8]. This was subsequently shown to be primarily due to reduction of the support accompanied by its migration or the migration of one of its suboxides onto the metal surface, thus causing decreased chemisorption capacity due to physical blockage of surface sites, rather than due to any significant electronic interaction [9,10]. However, this "SMSI" state, which is typically induced by a high-temperature reduction in H_2, was found to have a major synergistic effect on certain types of hydrogenation reactions, particularly those involving hydrogenation of a carbonyl bond [11–14], and this is attributed to the creation of new active sites at the metal-support interface due to the removal of oxygen atoms from the surface of the oxide lattice structure [13–16]. This latter situation is the one which is most appropriately designated as an MSI (Metal-Support Interaction) effect.

References

1. a) *Pure Appl. Chem.* 45 (1976) 71.
   b) *Adv. Catal.* 26 (1976) 351.
2. M. Boudart and G. Djéga-Mariadassou, "Kinetics of Heterogeneous Catalytic Reactions", Princeton University Press, Princeton, NJ, 1984.
3. M. Boudart, *Adv. Catal.* 20 (1969) 153.
4. G. Djéga-Mariadassou and M. Boudart, *J. Catal.* 216 (2003) 89.
5. M. Boudart, "Kinetics of Chemical Processes", Prentice-Hall, Englewood Cliffs, NJ, 1968.
6. C. G. Hill, "An Introduction to Chemical Engineering Kinetics & Reactor Design", J. Wiley, NY, 1977.
7. a) B. C. Gates, *Chem. Rev.* 95 (1995) 511.
   b) J. Guzman and B. C. Gates, *J. Chem. Soc. Dalton Trans.* (2003) 3303.
8. a) S. J. Tauster, S. C. Fung and R. L. Garten, *J. Am. Chem. Soc.* 100 (1978) 170.
   b) R. T. K. Baker, E. B. Prestridge and R. L. Garten, *J. Catal.* 56 (1979) 390; *J. Catal.* 59 (1979) 293.
9. a) R.T. K. Baker, E. B. Prestridge and R. L. Garten, *J. Catal.* 79 (1983) 348.
   b) S. J. Tauster, *Acct. Chem. Res.* 20 (1987) 389.
10. G. L. Haller and D. E. Resasco, *Adv. Catal.* 36 (1989) 173.
11. M. A. Vannice and R. L. Garten, *J. Catal.* 56 (1979) 236.
12. M. A. Vannice, *J. Catal.* 74 (1982) 199.
13. M. A. Vannice, *J. Molec. Catal.* 59 (1990) 165.
14. M. A. Vannice, *Topics in Catalysis* 4 (1997) 241.
15. R. Burch and A. R. Flambard, *J. Catal.* 78 (1982) 389.
16. M. A. Vannice and C. Sudhakar, *J. Phys. Chem.* 88 (1984) 2429.

3
Catalyst Characterization

In a proper catalytic study, as much as possible should be learned about the physical properties of the catalyst employed. These properties include the total surface area of the catalyst, the average pore size and/or the pore size distribution, the metal surface area and average metal crystallite size, especially in supported metal catalysts, as well as the metal weight loading of the latter. From this information and the packing densities of different crystallographic planes, one can calculate the dispersion of the metal, D_M, defined as the fraction of the total amount of metal atoms, N_{M_t}, that exist as surface atoms, N_{M_s}, thus,

$$D_M = N_{M_s}/N_{M_t} \tag{3.1}$$

as defined in Chapter 2.12. This term is also referred to as the metal fraction exposed [1], and it can be determined by utilizing an appropriate chemisorption technique for which the adsorption stoichiometry is known, thus the number of surface atoms is counted directly. This latter information, which is required to calculate dispersion, is also necessary to calculate a turnover frequency (TOF) and to examine crystallite size effects in a catalytic reaction, whereas the former type of information about pore size is needed to check for mass transfer limitations, as discussed in Chapter 4.

The aspects of catalyst characterization have been discussed in detail elsewhere, for example, see references [2–6], and the numerous methods, some of which are restricted to specific elements and isotopes, that can supply the above knowledge have been tabulated. However, a review should be useful here of only the methods that are most readily available to the general catalytic researcher, such as TEM (Transmission Electron Microscopy) including SEM (Scanning Electron Microscopy) and HREM (High Resolution Electron Microscopy), XRD (X-ray Diffraction), physisorption and chemisorption. Because of the unique information it can provide, EXAFS (Extended X-ray Absorption Fine Structure) is also discussed briefly.

A typical supported metal catalyst pellet, i.e., one with metal crystallites dispersed throughout a porous support with a high surface area, such as SiO_2, Al_2O_3 or carbon, can be depicted as shown in Figure 3.1. The size of

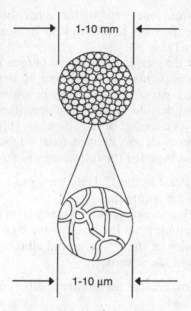

1-10 mm

1-10 μm

FIGURE 3.1. Schematic representation of a typical catalyst pellet comprised of small porous particles.

the catalyst pellet used in a commercial reactor can typically range from 1–10 mm, and this granule is usually composed of smaller particles frequently on the order of 1 micron ($1\mu = 10^{-6}$m $= 10^{4}$Å) that are held together by a binder added during the catalyst preparation process. These smaller particles are usually porous and contain micropores (pore diameter $\leq$ 2 nm) and mesopores (2 nm < pore diameter $\leq$ 50 nm) [1]. Macropores (pore diameter > 50 nm) can also exist which can be dependent on the size and packing density of these small particles. Metal particles are routinely distributed throughout the entire pore structure, as indicated by the dark particles in Figure 3.1, although techniques exist to prepare metal catalysts with the metal component located only at the center of each small support particle (cherry model) or only at the external surface of each support particle (eggshell model) [7].

3.1   Total (BET) Surface Area

It is of course important to know the total surface area, the pore volume and the pore size distribution of such porous materials. The first property is obtained by the nonselective physical adsorption of an appropriate adsorbate, typically dinitrogen (N_2), at liquid nitrogen temperature (ca. 77 K) although many other adsorbates have also been used [8–10]. The distinction between physical and chemical adsorption is not always clearly defined as there is a transition from one state to the other, but certain identifying characteristics can be specified to distinguish between the two forms of adsorption, as listed in Table 3.1.

A major advance in the characterization of porous materials, in general, and solid catalysts, in particular, was achieved by Brunauer, Emmett and Teller in their landmark paper in which they proposed a model for multi-layer physisorption and derived an equation describing it that allowed the calculation of monolayer coverage of the adsorbate [11]. This equation, now known as the BET equation, was derived from a model that extended the Langmuir isotherm and included the following assumptions:

a) Each adsorbed molecule in the 1st layer serves as a site for the 2nd layer (lateral interactions are ignored).
b) The rate of adsorption (condensation) on any layer (x) equals the rate of desorption (evaporation) from the layer above it (x + 1).
c) The heat of adsorption of the 2nd layer and all those above it equals the heat of liquefaction of the adsorbate.

All layers were summed, and for an infinite number of layers the following BET equation is obtained:

$$\frac{P}{V(P_o - P)} = \left[\frac{(C-1)}{V_m C}\right]\frac{P}{P_o} + \frac{1}{V_m C} \tag{3.2}$$

where V = total volume adsorbed (STP) at pressure P, V_m is the volume adsorbed (STP) at monolayer coverage, P_o is the saturation vapor pressure of the adsorbate gas (or vapor), and, to a good approximation:

$$C = e^{(q_1 - q_L)/RT} \tag{3.3}$$

where q_1 is the heat of adsorption in the 1st monolayer and q_L is the heat of liquefaction (condensation) of the adsorbate. With a minor rearrangement of equation 3.2, one gets

$$\frac{(P/P_o)}{V(1 - P/P_o)} = \frac{(C-1)}{V_m C}(P/P_o) + \frac{1}{V_m C} \tag{3.4}$$

TABLE 3.1. Differing characteristics between physical and chemical adsorption.

Physisorption	Chemisorption
1. No electron transfer – no bonding, van der Waals forces, weak interaction	Electron transfer – chemical bonds formed, strong interaction
2. Heat of adsorption, Q_p, low – similar to heat of liquefaction	Heat of adsorption, Q_c, high, typically $\geq$ 10 kcal/mole
3. Occurs near or below boiling point of adsorbate	Occurs at temperatures far above boiling point
4. Non-activated	May be activated
5. Non-specific	Very specific, depends on surface
6. Multilayer	Monolayer

and a plot of the left-hand side vs. P/P_o should give a linear plot with a slope $= (C-1)/V_mC$ and an intercept $= 1/V_mC$, from which V_m and C can be determined. The preferred range of P/P_o for best results is 0.05 to 0.4 [8]. With N_2 and many other adsorbates $C \gg 1$, thus to a good approximation the slope equals $1/V_m$. Based on values of C and q_L, heats of physisorption represented by q_1 are typically below 10 kcal/mole [9]. A specific surface area, A, is then easily calculated from the V_m values; for example,

$$A \ (m^2 g^{-1}) = \left(V_m, \frac{cm^3 STP}{g} \right) \left(\frac{6.023 \times 10^{23} \ molecules}{21400 \ cm^3 STP} \right)$$
$$\left(\frac{\text{cross-sectional area, } m^2}{\text{molecule}} \right) \tag{3.5}$$

and for N_2 the usual cross-sectional area is $16.2 \, \text{Å}^2$ ($1\text{Å}^2 = 10^{-20} \, m^2$) [10]. Note that V and V_m can easily be replaced by n, the adsorbate uptake, and n_m, the monolayer adsorbate uptake, respectively.

3.2 Pore Volume and Pore Size Distribution

3.2.1 Hg Porosimetry Method

A simple model for the pore volume or the void space in a porous material is to assume it to be composed of a collection of cylindrical pores of radius r. Then a volume of a liquid that does not wet the pore wall surfaces can be forced under pressure to fill the void space. This liquid is invariably mercury because it has a high surface tension, thus the Hg penetration (or porosimetry) method is used to determine pore volumes and the pore size distribution of larger pores, i.e., those with radii larger than about 10 nm. The relationship between pore size and applied pressure, P_{ap}, is obtained by a force balance, that is, the force due to surface tension is equated to the applied force:

$$-2\pi \, r\sigma \cos \theta = \pi r^2 P_{ap} \tag{3.6}$$

where σ is the surface tension and θ is the contact angle between the liquid and the pore wall. These respective values for Hg are typically 480 dyne/cm (or 0.48 N/m) in air at 293 K and 140°, thus using these values and rearranging equation 3.5 gives:

$$r(\text{Å}) = \frac{1.07 \times 10^6}{P_{ap}(\text{psia})} \tag{3.7a}$$

or

$$r(\text{nm}) = \frac{7.37 \times 10^8}{P_{ap}(\text{Pa})} \tag{3.7b}$$

The constant in this equation is slightly different from older values [2]. From this equation it can be easily seen that pressures near 10^4psia, which is the physical limit of most experimental systems, are required to fill pores 200 Å in diameter, hence the lower limit of $r = 10$ nm mentioned above.

3.2.2 N_2 Desorption Method

To determine the distribution of pores with diameters smaller than 20 nm, a nitrogen desorption technique is employed which utilizes the Kelvin equation to relate the pore radius to the ambient pressure. The porous material is exposed to high pressures of N_2 such that $P/P_o \rightarrow 1$ and the void space is assumed to be filled with condensed N_2, then the pressure is lowered in increments to obtain a desorption isotherm. The vapor pressure of a liquid in a capillary depends on the radius of curvature, but in pores larger than 20 nm in diameter the radius of curvature has little effect on the vapor pressure; however, this is of little importance because this region is overlapped by the Hg penetration method.

The Kelvin equation shows that the smaller the pore radius, the lower the vapor pressure, P, in the pore:

$$r - \delta = 2\sigma V_\ell \cos \theta / RT \, \ell n(P/P_o) \tag{3.8}$$

where σ, θ and P_o are defined as before, V_ℓ is the molal volume of the condensed liquid and δ is the thickness of the monolayer adsorbed on the pore wall. This latter correction is necessary because equation 3.7 applies only to condensed molecules and not to adsorbed molecules, which can be affected by their interaction with the surface. Also, for a liquid such as condensed N_2 that wets the pore surface, in contrast to Hg, the wetting angle is near zero so that $\cos \theta$ is essentially unity. Then, based on the analysis of Wheeler for N_2 [12]:

$$r - \delta(\overset{\circ}{A}) = 9.52 / \log (P_o/P) \tag{3.9}$$

and

$$\delta(\overset{\circ}{A}) = 7.34 \, [\log (P_o/P)]^{-1/3} \tag{3.10}$$

3.2.3   Overall Pore Size Distribution

By combining these two techniques, the cumulative pore volume, V_p, vs. pore radius, r, can now be obtained. As one is typically more interested in the distribution of pores, i.e., dV_p/dr, this quantity is obtained from the slope at different positions on a smooth curve drawn through the data points of the cumulative volume, and it is usually plotted as $dV_p/d \ln r$ vs. $\ln r$, as shown by the example for the Hg penetration method provided in Figure 3.2 [13].

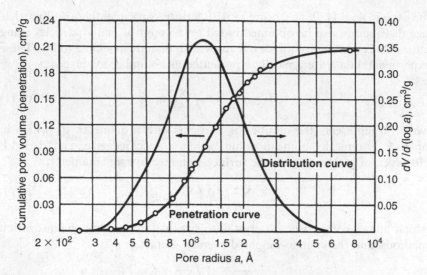

FIGURE 3.2. Pore volume distribution in a UO_2 pellet. (Reproduced from ref. 13, copyright © 1981, with permission of the McGraw-Hill Companies)

One important aspect of the utilization of the latter method is the appearance of an artifact due to the tensile strength of the adsorbate which causes a false maximum to frequently occur between r values of 1.7–2.0 nm when N_2 is used [10]. Although this artifact has been misinterpreted in numerous papers and reported as an actual sharp distribution of small pores, one must not be misled by it.

3.3 Metal Surface Area, Crystallite Size, and Dispersion

Many catalytic studies, perhaps even a majority, have involved metallic systems, either unsupported or supported on a high surface area substrate which is frequently inert in the reaction of interest. Thus the reaction rate is dependent on the specific surface area (m^2g^{-1}) of the metal, not only because the total number of active sites can vary, but also because the average metal crystallite size is dependent on this value and some reactions, now termed structure-sensitive [14], have areal rates (and TOFs) that are dependent on crystallite size [14,15]. Consequently, it is of utmost importance to measure the metal surface areas in these catalysts and calculate metal dispersions and crystallite sizes based on this information. The three most general approaches to accomplish this involve TEM (SEM), XRD, and chemisorption methods.

3.3.1 Transmission Electron Microscopy (TEM)

Using appropriately prepared samples, TEM in its various modes (SEM, HREM) can routinely resolve metal crystallites and clusters down to sizes at the atomic level [4,5], i.e., below 5 Å (0.5 nm); consequently, a full particle size distribution can be obtained based on a count of many particles using automated counting techniques, if accessible, that are available today. These experimental data are typically represented as a number average:

$$\bar{d}_n = \sum_i n_i d_i \bigg/ \sum_i n_i \qquad (3.11)$$

where n_i represents the number of particles with a diameter, d_i, which in practice is typically a mean value within a specified size, i.e., $d_i \pm \Delta d$. However, of greater use are the surface-weighted average diameter:

$$\bar{d}_s = \sum_i n_i d_i^3 \bigg/ \sum_i n_i d_i^2 \qquad (3.12)$$

which allows comparison to crystallite sizes determined by chemisorption methods, and the volume-weighted average diameter:

$$\bar{d}_v = \sum_i n_i d_i^4 \bigg/ \sum_i n_i d_i^3 \qquad (3.13)$$

which allows comparison to crystallite sizes obtained from XRD analysis or magnetic measurements.

A narrow crystallite size distribution will result in similar values for equations 3.12 and 3.13, and these are frequently in good agreement with surface- and volume-weighted averages determined by chemisorption and XRD measurements, as shown in Table 3.2 [3]. The broader the distribution, the greater the disparities among the different $\bar{d}$ values. A bimodal crystallite size distribution can also cause large discrepancies among these differently weighted $\bar{d}$ values, regardless of the method used for characterization.

3.3.2   X-Ray Techniques

3.3.2.1   Line Broadening of X-Ray Diffraction (XRD) Peaks

X-ray diffraction patterns can be rapidly obtained with today's computerized diffractometers, and they are very informative in regard to typical dispersed metal catalysts. The high-surface-area support material usually has quite broad diffraction peaks because in its amorphous state it has limited long-range, translational order. In contrast, diffraction peaks for

TABLE 3.2. Comparison of average Pt crystallite diameters for a Pt/SiO$_2$ catalyst. (Reprinted from Ref. 3b, copyright © 1962, with permission from Elsevier)

Type of diameter	Average diameter (Å)		
	TEM	XRD	H$_2$ chemisorption
d_n (Eq. 3.11)	28.5	–	–
d_s (Eq. 3.12)	30.5	–	34.4
d_v (Eq. 3.13)	31.5	37.9	–

reduced metal crystallites are frequently much sharper; however, these peaks also broaden as the crystallites become smaller because fewer planes exist to give rise to Bragg diffraction. Thus this property can be used to calculate metal crystallite size over an appropriate size range in which the peaks are neither too sharp so they cannot be distinguished from the instrumental line broadening, nor too broad so they cannot be discerned from the background spectrum of the support. A reasonable size range for most applications is 3–50 nm, and the average size obtained is a volume-weighted average.

This relationship is acquired using the Scherrer equation for XRD line broadening [16]:

$$d_v = K\lambda/(\beta \cos \theta) \tag{3.14}$$

where d_v is the volume-weighted crystallite diameter, λ is the radiation wavelength (for the most common source, CuK_α radiation, the average value for λ is 0.15418 nm), β is the line broadening of a particular peak due to the crystallite size, K is a constant that is typically ≈ 0.9, and θ is the angle of the diffraction peak. Line broadening due to the diffractometer itself has to be determined vs. θ (or 2θ, as data are acquired as intensity vs. 2θ) using a well defined crystalline material, and β must then be corrected for this. A common way to do this is to use Warren's correction for β [16]:

$$\beta = (B^2 - b^2)^{1/2} \tag{3.15}$$

where B is the peak width at half height (in radians) and b is the instrumental broadening (in radians). When possible, it is beneficial to calculate d_v values for different diffraction peaks and check for consistency.

Also, most diffractometers are not set up to obtain patterns with samples under a controlled environment, thus nearly all XRD data are obtained after exposure to air. One must ascertain that such exposure is not going to alter the reduced state of the metal significantly, which could change the diffraction pattern, and even the formation of a thin oxide overlayer on a small metal crystallite could alter the metal peak width observed. Frequently, some passivation step should be used to prevent any significant oxidation, especially with small metal crystallites that are prone to oxidation (Fe and Ni, for example). Alternatively, leak-proof XRD sample holders can be made, and reduced samples can be loaded in a dry box prior to placement in the X-ray diffractometer [17].

It is worth mentioning that another x-ray technique – SAXS (Small-Angle X-ray Scattering) – exists which can be applied to characterize very small (<6 nm) metal crystallites dispersed in a porous matrix. However, it is more difficult to use as it requires that the pore volume be eliminated, either by crushing the sample at extremely high pressures or by filling the pore volume with a liquid possessing an electron density equal to that of the support [3,4]. This constraint has inhibited its common usage.

3.3.2.2 Extended X-Ray Absorption Fine Structure (EXAFS)

The characterization techniques discussed previously in this book have been purposely restricted to those that should be readily available to most investigators in academic or industrial environments. The one exception is this one – EXAFS – which is due to its specific advantages when applied to supported catalysts, especially highly dispersed metal systems [18–20]. Although the collection of EXAFS data typically requires a high-energy synchrotron radiation source, which gives short data acquisition times, a number of such storage rings now exist and their usage is available to various researchers.

XRD requires a structure periodicity that extends more than 2–3 nm to produce a discernible diffraction pattern amenable to analyses using the Bragg equation and, subsequently, the Scherrer equation (eq. 3.14). In contrast, this size limitation does not apply to X-ray absorption spectroscopy in which inner atomic shells are ionized and photoelectrons are emitted. The fine structure is a consequence of the interference that develops between the outgoing photoelectron wave and the portion that is backscattered from neighboring atoms and, as a result, it is a direct probe of the local atomic environment around the absorbing atom. Consequently, EXAFS is particularly well suited for catalyst structure and adsorption studies not only because it can determine the short-range order around a particular atomic species in both periodic and nonperiodic surroundings, but it can also measure bond lengths and their variation around an average value. This information is independent of particle size and does not require that the crystal structure be known. Another significant advantage of this technique is that such structural information can be obtained in situ under high pressure and/or reaction conditions, in contrast to the vacuum requirements associated with electron probes.

As mentioned above, EXAFS is an interference phenomenon which can be visualized as resulting from two basic processes, i.e., a photoelectric effect due to absorption of an x-ray photon and an electron diffraction process in which the electron source and detector are present in the atom absorbing the photon. Clearly, multiple-scattering events can occur and interpretation of the data is not straightforward; however, a radial distribution function for short-range order around the absorbing atom can be obtained. The theory utilized to acquire this radial distribution function and the distances involved is complex and will not be addressed here; however, it is discussed elsewhere in detail [19,20]. The utilization of EXAFS has grown significantly and it has been applied to a number of problems related to catalysts, such as: 1) the effect of pretreatment on structure, 2) variations in metal dispersion due to preparation or aging, 3) bimetallic clusters, 4) design of metal surfaces, 5) surface compositions and 6) metal-support interactions [18,19].

3.3.3 *Magnetic Measurements*

With ferromagnetic metals that exhibit superparamagnetism, magnetic measurements can also be used to calculate particle size [2,4,21]. For example, the low-field and high-field Langevin equations were used to analyze the magnetization behavior of Fe particles dispersed on carbon, and the results indicated the presence of 2–4 nm Fe crystallites, a size range consistent with previous measurements [22].

3.3.4 *Chemisorption Methods*

The most sensitive techniques to count metal surface atoms are those involving selective chemisorption (or titration) methods because all surface atoms, independent of crystallite size, are probed at the molecular level. Crystallite sizes based on this approach are surface-weighted average sizes. For such techniques to be successful, the adsorption stoichiometry on the metal surface must be known, at least to a good approximation, and uptakes must be correctable for adsorption on the support, when present. For a specified adsorbate, this stoichiometry can depend on the metal, the distribution of crystal planes exposed (i.e., the surface coordination numbers), crystallite size, the temperature and other experimental factors; regardless, with the use of well-defined experimental procedures, quite reproducible results can be obtained in different laboratories for many different metal catalysts and, in some cases, nonmetallic systems. This approach is readily applicable to supported metal systems, especially those containing a large fraction of very small (0.3–2 nm) particles which are difficult to detect by TEM and XRD methods, because strong irreversible chemisorption typically occurs on the reduced metal surface while weak reversible adsorption exists on the support. Therefore, the former uptakes can be modeled by a Langmuir isotherm and should be measured in a high-pressure regime where saturation coverage is obtained on the metal, whereas in this same pressure regime weak reversible adsorption on the support exists in the Henry's Law region (see section 5), where coverage is low and is directly proportional to the adsorbate pressure.

3.3.4.1 H_2 Chemisorption

The chemisorption of hydrogen is the most widely used method and it has been studied on all the Group VIII metals [3,4,23]. Adsorption is routinely dissociative and is applicable around 300 K for all these metals except Fe, on which surface it frequently exhibits activated adsorption [24], especially on very small Fe crystallites [25], thus it is not applicable to this metal. However, it is widely used with the other metals [26], and such a system is represented in Figure 3.3 for a Pt/Al_2O_3 catalyst, but other Group VIII metals and different supports, such as SiO_2, TiO_2, molecular sieves, and carbon, would provide similar examples. Studies of H_2 adsorption on

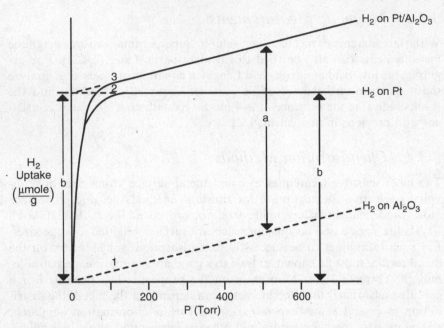

FIGURE 3.3. Representative adsorption isotherms for Pt dispersed on Al_2O_3: 1) H_2 adsorption on Al_2O_3 in Henry's law region, 2) H chemisorption on Pt, 3) Total H_2 chemisorption on Pt/Al_2O_3 catalyst.

unsupported Pt surfaces have shown that it is dissociative, and at temperatures easily utilized in experiments (ca. 300 K) saturation coverages are reached at H_2 pressures above 50–100 Torr (760 Torr = 101.3 kPa = 1 atm) that give an adsorption stoichiometry very close to $H_{ad}/Pt_s = 1$, where H_{ad} is an adsorbed H atom and Pt_s is a surface Pt atom [27,28]. This stoichiometry was determined on bulk Pt, i.e., large unsupported Pt crystallites, and it represents a good approximation for most Pt surfaces; however, it must be recognized that adsorption on metal clusters and very small (1–3 nm) crystallites with low surface coordination numbers can allow this ratio to increase, possibly due to a decrease in steric constraints, and H_{ad}/Pt_s ratios have been reported that are greater than unity and can approach 2 [23,26]. Regardless, H_2 chemisorption on a typical unsupported Pt surface can be represented well by the Langmuir-type isotherm shown in Figure 3.3, which for simplicity is represented as H_2 uptake (μmole g^{-1}) vs. H_2 pressure, and saturation coverage is achieved at pressures above several hundred Torr. In this latter pressure regime, H_2 adsorption on pure Al_2O_3 is weak and reversible and described by Henry's Law (Line 1), while the total uptake on the Pt/Al_2O_3 catalyst in Figure 3.3 is given by Line 3 and it represents the sum of both contributions. The difference, a, between lines 1 and 2 represents the adsorption associated with only the Pt crystallites (line 2), and it is used to count surface Pt atoms using the appropriate stoichiometry, i.e.,

(3.1) $H_2 + 2Pt_s \longrightarrow 2Pt_s - H$

This stoichiometry is usually applicable to most of the Group VIII metals at temperatures around 300 K [26], i.e.,

(3.2) $H_2 + 2M_s \longrightarrow 2M_s - H$

For many adsorbate/support systems Henry's Law is obeyed and, as a consequence, this procedure is simplified for the experimenter because Line 3 in Figure 3.3 can simply be extrapolated to its zero-pressure intercept to obtain b, which is equal to a, thus eliminating the need to obtain a second isotherm. This latter procedure does not represent the gas uptake at zero pressure, as sometimes mis-stated, rather it represents subtraction of adsorption on the support at "saturation" coverage on the metal.

At temperatures near 300 K (± 50K), H_2 chemisorption may be well represented by a Langmuir isotherm, but the heat of adsorption routinely decreases with coverage until it becomes low enough (ca. 10 kcal mole^{-1}) that sufficient thermal energy exists at these temperatures to allow some rapid desorption to occur. Therefore, if the sample is evacuated following adsorption at 300 K and a second isotherm for H_2 adsorption on the Pt (or other metal) is measured, some reversible H_2 adsorption will be detected. A valid question is then: "Which of the two H_2 uptake values – the total uptake represented by Line 2 or the irreversible uptake only (the difference between the two isotherms) – corresponds most closely to the stoichiometry given by reactions 3.1 and 3.2?" The answer to this has not been unambiguously ascertained – certainly for larger Pt particles the former is preferred because it has been checked by BET measurements [27]; however, for small, dispersed Pt crystallites the former value can easily overestimate dispersion whereas the latter value very likely underestimates it. Keep in mind that the dispersion, D, cannot exceed unity even though the ratio of H atoms adsorbed to the total number of metal atoms can, i.e., 2(mole H_2 adsorbed per g cat)N_{Av}/N_{M_t}(per g cat) > 1. Consequently, H_{ad}/M_t ratios in excess of unity are a strong indication of very small metal particles (with $D \cong 1.0$), and this information can be coupled with the "irreversible" H adsorption which would place a lower limit on D. In evaluating these uptakes, complications such as hydrogen spillover [29,30] must also be considered and, if possible, eliminated.

Special attention must be given to metals which can form a hydride phase [31], particularly Pd, which is one of the most commonly used Group VIII metal catalysts. The β-hydride phase in Pd becomes thermodynamically favorable at temperatures near 300 K once the H_2 pressure increases above 1–2 Torr, and hydrogen absorption occurs to give a bulk hydride stoichiometry near PdH$_{0.6}$ [31,32]. Consequently, if H_2 chemisorption is used to measure Pd surface area, experimental procedures must be chosen to either avoid bulk hydride formation [33,34] or to allow uptakes to be corrected for it [35,36]. This can be accomplished by preventing β-phase hydride formation by either measuring H coverages at 300 K and at H_2 pressures below 1 Torr or by measuring coverages at 373 K and much higher H_2 pressures

(200–350 Torr) [32–36]. Alternatively, higher H_2 pressures at 300 K can be used to both form the hydride and saturate the Pd surface with H atoms, then an evacuation step is used to rapidly decompose the bulk hydride, and this is followed by obtaining a second isotherm. Similar to the situation in Figure 3.3, the difference, a, represents the irreversible H adsorption on the Pd surface. This approach may be preferred because it provides additional information about the Pd crystallites, i.e., once the surface Pd_s atoms are counted by the irreversible uptake, the remainder of the atoms can be attributed to bulk (Pd_b) atoms, i.e., $Pd_b = Pd_t - Pd_s$, and the apparent bulk hydride ratio can be determined [32]. Values near $PdH_{0.6}$ are typically attained with large clean Pd crystallites, but on small Pd crystallites this apparent hydride ratio can become larger than 0.6 because reversible chemisorption on the Pd_s atoms can dominate the second isotherm as the Pd dispersion approaches unity. Consequently, valuable information can be obtained regarding surface cleanliness and metal-support interactions (MSI) [32,37]. An example of such an effort is provided by Illustration 3.1.

Illustration 3.1 – Determination of Pd Dispersion and Crystallite Size by Chemisorption Methods

A family of catalysts was studied by Chou and Vannice to examine the influence of crystallite size on H_2 uptakes and heats of adsorption on Pd at 300 K [32]. Dispersions and crystallite sizes were determined by several chemisorption methods at 300 K, including CO and O_2 adsorption [42,43] as well as H_2 chemisorption, and in some cases H_2 uptakes at 373 K were also measured. Using the irreversible hydrogen uptake at 300 K to count Pd_s atoms, apparent bulk hydride ratios could be calculated by subtracting surface Pd atoms from total Pd atoms, i.e., $Pd_b = Pd_T - Pd_s$. Some represen- tatitive adsorption isotherms are shown in Figure 3.4 for Pd powder, a rather poorly dispersed (D = 0.17) $Pd/SiO_2-Al_2O_3$ catalyst, and a more highly dispersed (D = 0.69) Pd/SiO_2 catalyst. Table 3.3 lists the uptakes of the three gases, the dispersions and average crystallite sizes based on the irreversible H uptake, and the calculated β-phase hydride ratios.

First of all, note that hydrogen adsorption on the surface of the Pd powder is noticeably higher than that for the other two adsorbates, although the coverages of CO molecules and O atoms are very similar (4.5 vs. 5.0 $\mu molg^{-1}$; however, the bulk β-hydride ratio of 0.66 is almost exactly that expected from phase diagrams. Larger Pd crystallites tend to give bulk hydride ratios similar to that for unsupported Pd, and large deviations from this ratio can indicate problems with a catalyst [37]. However, for very small Pd crystallites these ratios are significantly higher, which is undoubtedly due to the inclusion of reversible adsorption that now constitutes a large fraction of this uptake. Figure 3.4(b) shows that the transition from the α- to the

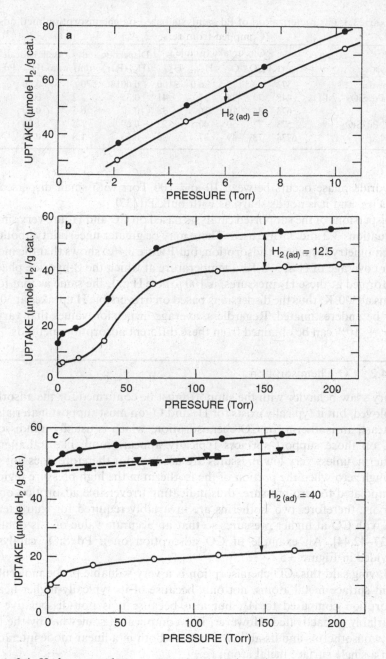

FIGURE 3.4. Hydrogen and oxygen uptakes on Pd catalysts – a) Unsupported Pd powder, b) 1.95% Pd/SiO$_2$ – Al$_2$O$_3$ (T$_{reduction}$ = 673 K), c) 1.23% Pd/SiO$_2$: H$_2$ uptakes at 300 K (●, ○); H$_2$ uptake at 373 K (▼); O$_2$ uptake at 300 K (µmole O$_2$/g) (■). Solid symbols: total uptake; open symbols: reversible uptake. (Reprinted from ref. 32, copyright © 1987, with permission from Elsevier)

TABLE 3.3. Characterization of Pd catalysts based on chemisorption methods. (Compiled from refs. 32, 42)

Catalyst	T_{red} (K)	CO	O_2	$H_{2\ irr}$	$H_{2\ rev}$	Dispersion (H_{irr}/Pd_T)	$\bar{d}$ (nm)	Bulk hydride ratio (H_{rev}/Pd_b)
Pd powder	573	4.5	2.5	6.0	3100	0.0013	870	0.66
1.9% Pd/SiO$_2$–Al$_2$O$_3$	448	70	24	33	41	0.35	3.2	0.69
	673	34	19	16	38	0.17	6.5	0.50
1.23% Pd/SiO$_2$	573	77	50	40	22	0.69	1.6	1.24
	6734	78	49	37	20	0.63	1.8	0.94

β-hydride phase occurs between 10 and 100 Torr with small dispersed Pd particles, and it is not as sharp as with bulk Pd [37].

Dispersions of the supported catalysts based on CO and H$_2$ are very similar if equations 3.2 and 3.3 are used. There may be greater uncertainty about the stoichiometry for oxygen adsorption, but Figure 3.4(c) shows that the monolayer coverage of H at 373 K, a temperature at which the β-hydride phase is not formed at these H$_2$ pressures, is 100 μmole H g^{-1}, the same as that for O atoms at 300 K, thus the dispersions based on irreversible H uptakes at 300 K may be underestimated. Regardless, average dispersion values that vary by only ± 10% can be obtained from these different adsorption methods.

3.3.4.2 CO Chemisorption

Henry's law behavior with the support must be confirmed for the adsorbate employed, but it typically exists for H$_2$ and O$_2$ on most support materials. In contrast, an isotherm for CO chemisorption, which is usually nondissociative, on those support surfaces typically appears more like a Langmuir isotherm, unless very low pressures are used, and it therefore does not pass through zero when the portion of the isotherm in the high pressure region is extrapolated to zero pressure, thus indicating irreversible adsorption on the support; therefore, two isotherms are invariably required for characterization with CO at higher pressures so that an accurate value of a is obtained [23,37–42,44]. An example of CO adsorption on a Pd/SiO$_2$ catalyst is provided in Figure 3.5

Having said this, CO chemisorption is a very valuable probe molecule to count surface metal atoms, not only because of its typically higher heat of adsorption compared to H$_2$, but also because it is non-dissociative and invariably nonactivated. However, it is complicated somewhat by the need for two isotherms and its ability to adsorb both in a linear mode interacting with a single surface metal atom, i.e.,

(3.3)
$$CO + M_s \longrightarrow M_s\overset{\displaystyle O}{\underset{\|}{C}}{}^{\nearrow}$$

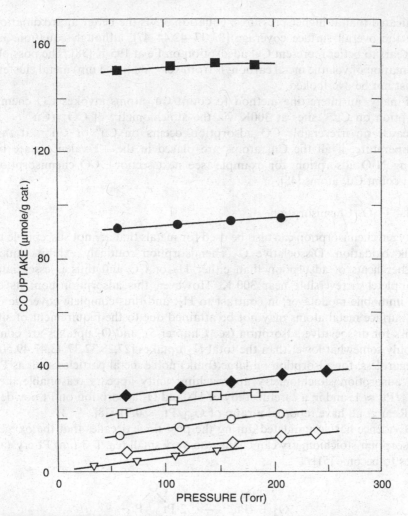

FIGURE 3.5. CO adsorption isotherms at 300 K on pure SiO_2 and Pd/SiO_2 catalysts – 2.10% Pd/SiO_2 (■, □); 1.23% Pd/SiO_2 (●,○); 0.48% Pd/SiO_2 (◆, ◇); pure SiO_2 (▽): Solid symbols: total CO uptake; open symbols: reversible CO uptake. (Reprinted from ref. 42, copyright © 1987, with permission from Elsevier)

as well as in a bridged mode in which it binds to 2 surface atoms:

$$
\begin{array}{c}
O \\
\parallel \\
C \\
\diagup \ \diagdown
\end{array}
$$

(3.4) $CO + 2M_s \longrightarrow M_s \quad M_s$

Despite this uncertainty in stoichiometry, numerous comparisons of CO uptakes versus H and O uptakes on many Group VIII metal surfaces have indicated that at higher pressures, equation 3.3 is the better approximation for the overall surface coverage [23,37–42,44–47], although equation 3.4 appears to better represent CO adsorption on Fe at 195 K [38]. The possible formation of volatile metal carbonyls from very small (< 1 nm) metal clusters must not be overlooked.

Finally, an interesting method to count Cu_s atoms invokes CO chemisorption on Cu^{+1} sites at 300K via the stoichiometry of $CO_{ad}/Cu^{+1} = 1$. Because no irreversible CO adsorption occurs on Cu^0 or Cu^{+2} at this temperature, if all the Cu_s atoms are placed in the +1 valence state by using N_2O adsorption, for example (see next section), CO chemisorption can count Cu_s atoms [48].

3.3.4.3 O_2 Chemisorption

Oxygen chemisorption can also be used for metals that are not susceptible to bulk oxidation. Dissociative O_2 chemisorption routinely exhibits much higher heats of adsorption than either H_2 or CO and thus is essentially completely irreversible near 300 K. However, this adsorption consists of an immobile monolayer, in contrast to H_2, and thus complete coverage of all surface metal atoms may not be attained due to the requirement of site pairs for dissociative adsorption (see Chapter 5), and O_2 uptakes are commonly somewhat lower than the total H_2 uptakes [27,28,32,37,43,47,49,50]. Regardless, for adsorption on large (bulk) noble metal particles such as Pt, an adsorption stoichiometry approaching unity appears reasonable, i.e., $O_{ad}/Pt_s \approx 1$, and in a careful study of H_2 and O_2 adsorption on Pt powder, O'Rear et al. have reported a ratio of $O_{ad}/Pt_s = 0.71$ [28].

Evidence has accumulated during the past three decades that the oxygen adsorption stoichiometry can change on very small (ca. 1–3 nm) Pt crystallites to become [51]:

$$(3.5) \qquad O_2 + 4Pt_s \longrightarrow 2 \, Pt_s \overset{O}{\diagup} \diagdown Pt_s$$

Thus there may be some uncertainty about the O_{ad}/Pt_s ratio (and the O_{ad}/M_s ratio, in general) unless additional information about the metal crystallite size is known. Regardless, in most situations the following assumption for surface stoichiometry should be satisfactory:

$$(3.6) \qquad O_2 + 2M_s \longrightarrow 2M_s - O$$

On some reduced metals, such as Cu and Ag, the chemisorption of H_2 or CO does not occur to a significant extent and the surface is not saturated, thus these molecules cannot be used to count surface atoms; however, oxygen chemisorption can be used if appropriate measures are taken so that bulk oxidation is avoided and a known adsorption stoichiometry is achieved. One route to achieve this involves dissociative N_2O adsorption on

Cu and Ag at temperatures near 363 K [48,52–56], and another route for Ag utilizes O_2 adsorption at 443 K [54–57]. With Cu, the adsorption stoichiometry is:

(3.7)
$$N_2O_{(g)} + 2Cu_s \longrightarrow Cu_s \overset{O}{\wedge} Cu_s + N_{2(g)}$$

and the O uptake is typically measured gravimetrically because there is essentially no change in the number of gas-phase molecules (only weak adsorption on the support) [48]. The adsorption stoichiometry is different with Ag [54–57], i.e.,

(3.8)
$$N_2O_{(g)} + Ag_s \longrightarrow Ag_s - O + N_2$$

This latter stoichiometry has been found recently to also be applicable to Pt [58].

3.3.4.4 H_2–O_2 Titration Techniques

One valuable chemisorption technique is the hydrogen-oxygen titration reaction. First proposed by Benson and Boudart for Pt catalysts [59], it represents the titration of an oxygen monolayer by hydrogen near 300 K and its stoichiometry for Pt was predicted based on equations 3.1 and 3.6, i.e.,

(3.9)
$$Pt_s - O + 3/2H_2 \longrightarrow Pt_s - H + H_2O_{(ad)}$$

This technique provided three advantages: it increased sensitivity by 3-fold, it eliminated a concern about oxygen contamination, and it was more applicable than H_2 chemisorption to used catalysts, which might have some contamination. The small amounts of H_2O formed were adsorbed by the support or by the walls of the cell if $T < 273$ K [27,28]. Subsequent studies indicated that for very highly dispersed Pt catalysts this titration stoichiometry could change from $O_{ad}: H_{ad}: H_{titr} = 1:1:3$ (Eq. 3.9) to 1:2:4 [60]. Wilson and Hall then showed that chemical equation 3.6 for large Pt crystallites shifts to chemical equation 3.5 for small Pt particles, which then gives a 1:2:4 ratio [51], i.e., the titration reaction becomes

(3.10)
$$Pt_s \overset{O}{\wedge} Pt_s + 2H_2 \longrightarrow 2Pt_s - H + H_2O_{ad}$$

If very highly dispersed Pt catalysts ($D \geq 0.4$) are known to exist, reaction 3.10 is probably more accurate, but even if reaction 3.9 is used in the absence of this knowledge, the variation in dispersion calculations is no more than 25%. More recently O'Rear et al. have reported a ratio of about 1:1.5:3.5 for clean Pt powder, and these authors have provided a very precise pretreatment to use prior to H_2 adsorption and titration measurements [28]. This titration technique has been extended to characterize Pd catalysts at 373 K [36] and Ag catalysts at 443 K [55–57].

3.3.5   Relationships Between Metal Dispersion, Surface Area, and Crystallite Size

Metal dispersion has been defined earlier by Equation 3.1. The metal content is routinely determined by various analytical methods to obtain the weight percentage [3] and hence the total number of metal atoms, N_{M_t} (mole g^{-1}), is known. Once a particular adsorption technique is chosen and used, the appropriate adsorption stoichiometry directly gives the number of surface metal atoms, N_{M_s} (mole g^{-1}). Metal surface areas can then be determined based on an average site density, n_s, where $n_s = 1/\left(\frac{\text{surface area (per g)}}{N_{Av} N_{M_s}}\right)$ and the denominator is frequently approximated by assuming appropriate amounts of the three low-index planes for that metal structure. For example, this would typically constitute equal quantities of the (111), (100) and (110) faces for an fcc metal such as Pt and their use gives a site density of $n_s = 1.24 \times 10^{15} Pt_s$ cm^{-2} [4] and the area per metal atom, A_M, is then $1/n_s = 8.06 \times 10^{-16}$ cm^2 [4]. If a specific geometric shape is now assumed, such as a sphere, then the volume occupied by a bulk metal atom is:

$$V_M = m/\rho N_{Av} \tag{3.16}$$

where m is the atomic mass, ρ is the bulk density, and N_{Av} is Avogadro's number, and this gives a value of $V_m = 15.1 \times 10^{-16}$ cm^3 for Pt. The relationship between dispersion and crystallite size is then

$$D_M = 0.6(V_M/A_M)\ d\ (nm) \tag{3.17}$$

which for Pt is $D_M = 1.1/d$, with d given in nm. The same equation is obtained if cubical geometry is assumed. Values for n_s, A_M, and V_M are listed elsewhere for other metals [4], and appropriate relationships between D_M and d can be determined, but the derived equations are similar to Equation 3.17.

In summary, there are readily available and easy to use XRD and chemisorption techniques that are applicable to metal catalysts, in particular, and to other catalysts, in general. If available, TEM can add additional information. Chemisorption is the most sensitive method, and all kinetic studies of metal catalysts should be accompanied by a measurement of the metal surface area and dispersion via a standard adsorption procedure. For nonmetallic catalysts, adsorption sites can still be counted in many situations by finding the appropriate region of temperature and pressure to measure adsorption of one of the reactants, and some (or all) of these sites would be expected to be active sites under reaction conditions. Examples of such efforts have been reported for N_2O, NO and O_2 adsorption on Mn_2O_3 and

Mn_3O_4 catalysts used for N_2O and NO decomposition [61,62] and for NO and O_2 adsorption on La_2O_3 and other rare earth oxide catalysts used for the decomposition of NO and the reduction of NO with CH_4 and other gases [63–66].

In addition, in some reactions catalyzed by solid acids, such as zeolites, Haag and coworkers have shown a direct correlation between the concentration of surface acid sites and activity [67]. Consequently, it appears that each acid site constitutes an active site, and such sites can be measured by appropriate adsorption methods [68–70].

References

1. a. *Pure Appl. Chem.* 45 (1976) 71.
   b. *Adv. Catal.* 26 (1976) 351.
2. "Experimental Methods in Catalytic Research", Vol. I, R. B. Anderson, Ed., Academic Press, NY, 1968.
3. a. R. L. Moss in "Experimental Methods in Catalytic Research", Vol. II, R. B. Anderson and R. T. Dawson, Eds., p. 43, Academic Press, NY, 1976.
   b. C. R. Adams, H. A. Benesi, R. M. Curtis and R. G. Meisenheimer, *J. Catal.* 1 (1962) 336.
4. G. Bergeret and P.Gallezot, in "Handbook of Heterogeneous Catalysis", G. Ertl, H. Knözinger and J. Weitkamp, Eds., Vol. 2, 439, Wiley-VCH, Weinheim, 1997.
5. J. M. Thomas and W. J. Thomas, "Principles and Practice of Heterogeneous Catalysis", VCH, Weinheim, 1997.
6. W. N. Delgass, G. L. Haller, R. Kellerman and J. H. Lunsford, "Spectroscopy in Heterogeneous Catalysis," Academic Press, NY, 1979.
7. A. Gavriilidis, A. Varma and M. Morbidelli, *Catal. Rev.-Sci. Eng.* 35 (1993) 399.
8. P. H. Emmett, in "Catalysis", P. H. Emmett, Ed., Vol. 1 p. 31, Reinhold, New York, 1954.
9. D. O. Hayward and B. M. W. Trapnell, "Chemisorption", Buttersworth, London, 1964.
10. S. J. Gregg and K. S. W. Sing, "Adsorption, Surface Area and Porosity", 2nd ed., Academic Press, London, 1991.
11. S. Brunauer, P. H. Emmett and E. Teller, *J. Am. Chem. Soc.* 60 (1938) 309.
12. A. Wheeler, in "Catalysis" P. H. Emmett, Ed., Vol. 2, p. 118, Reinhold, New York, 1955.
13. J. M. Smith, in "Chemical Engineering Kinetics", 3rd ed., McGraw-Hill, NY, 1981.
14. M. Boudart, *Adv. Catal. Relat. Subj.* 20 (1969) 153.
15. M. Che and C. O. Bennett, *Adv. Catal.* 36 (1989) 55.
16. H. P. Klug and L. E. Alexander, "X-Ray Diffraction Procedures", Wiley, NY, 1954.
17. A. A. Chen, M. A. Vannice, and J. Phillips, *J. Phys. Chem.* 91 (1987) 6257.
18. B. S. Clausen, H. Topsøe and R. Frahm, *Adv. Catal.*, 42 (1998) 315.
19. J. C. J. Bart and G. Vlaic, *Adv. Catal.* 35 (1987) 1.
20. D. C. Koningsberger, B. L. Mojet, G. E. van Dorssen and D. E. Ramaker, *Topics in Catalysis* 10 (2000) 143.

21. P. W. Selwood, "Chemisorption and Magnetization", Academic Press, NY, 1975.
22. N. Krishnankutty, L. N. Mulay and M. A. Vannice, *J. Phys. Chem.* 96 (1992) 9944.
23. J. H. Sinfelt, *Catal. Rev.* 3 (1969) 175.
24. P. H. Emmett and S. J. Brunauer, *J. Am. Chem. Soc.* 59 (1937) 319.
25. J. J. Venter and M. A. Vannice, *J. Phys. Chem.* 93 (1989) 4158.
26. C. H. Bartholomew, "Catalysis", Vol. 11, Chap. 3, Royal Soc. London, 1994.
27. M. A. Vannice, J. E. Benson and M. Boudart, *J. Catal.* 16 (1970) 348.
28. D. J. O'Rear, D. G. Löffler and M. Boudart, *J. Catal.* 121 (1990) 131.
29. M. Boudart, M. A. Vannice, J. E. Benson, *Z. Physik. Chem. N. F.* 64 (1969) 171.
30. W. C. Conner, Jr., G. M. Pajonk and S. J. Teichner, *Adv. Catal.* 34 (1986) 1.
31. W. M. Mueller, J. P. Blackledge and G. G. Libowitz, "Metal Hydrides", Academic Press, NY, 1968.
32. P. Chou and M. A. Vannice, *J. Catal.* 104 (1987) 1.
33. P. C. Aben, *J. Catal.*, 10 (1968) 224.
34. P. A. Sermon, *J. Catal.* 24 (1972) 460.
35. M. Boudart and H. S. Hwang, *J. Catal.* 39 (1975) 44.
36. J. E. Benson, H. S. Hwang and M. Boudart, *J. Catal.* 30 (1973) 146.
37. N. Krishnankutty and M. A. Vannice, *J. Catal.* 155 (1995) 312.
38. P. H. Emmett and S. Brunauer, *J. Am. Chem. Soc.* 59 (1937) 310; 57 (1935) 1553.
39. J. S. Smith, P. A. Thrower and M.A. Vannice, *J. Catal.* 68 (1981) 270.
40. M. A. Vannice, L. C. Hasselbring and B. Sen, *J. Catal.* 95 (1985) 57.
41. M. A. Vannice, L. C. Hasselbring and B. Sen, *J. Catal.* 97 (1986) 66.
42. P. Chou and M. A. Vannice, *J. Catal.* 104 (1987) 17.
43. P. Chou and M. A. Vannice, *J. Catal.* 105 (1987) 342.
44. B. Sen and M. A. Vannice, *J. Catal.* 130 (1991) 9.
45. R. F. Hicks, Q.-J. Yen and A. T. Bell, *J. Catal.* 89 (1984) 498.
46. P. B. Wells, *Appl. Catal.* 18 (1985) 259.
47. J. Freel, *J. Catal.* 25 (1972) 149.
48. A. Dandekar and M. A. Vannice, *J. Catal.* 178 (1998) 621.
49. M. B. Palmer, Jr. and M. A. Vannice, *J. Chem. Technol. & Biotech.* 30 (1980) 205.
50. B. Sen and M. A. Vannice, *J. Catal.* 129 (1991) 31.
51. G. R. Wilson and W. K. Hall, *J. Catal.* 17 (1970) 190.
52. R. M. Dell, F. S. Stone, and P. F. Tiley, *Trans. Faraday Soc.* 49 (1953) 195.
53. J. J. F. Scholten and J. A. Konvalinka, *Trans. Faraday Soc.* 65 (1969) 2465.
54. J. J. F. Scholten, J. A. Konvalinka and F. W. Beeckman, *J. Catal.* 28 (1973) 209.
55. S. R. Seyedmonir, D. E. Strohmeyer, G. L. Geoffroy, M. A. Vannice, H. W. Young and J. W. Linowski, *J. Catal.* 87 (1984) 424.
56. S. R. Seyedmonir, D. E. Strohmeyer, G. L. Geoffroy and M. A. Vannice, *Ads. Sci. & Tech.* 1 (1984) 253.
57. D. E. Strohmeyer, G. L. Geoffroy and M. A. Vannice, *Appl. Catal.* 7 (1983) 189.
58. M. H. Kim, J. R. Ebner, R. M. Friedman and M. A. Vannice, *J. Catal.* 204 (2001) 348.
59. J. E. Benson and M. Boudart, *J. Catal.* 4 (1965) 704.
60. D. E. Mears and R. C. Hansford, *J. Catal.* 9 (1967) 125.
61. T. Yamashita and M. A. Vannice, *J. Catal.* 161 (1996) 254.
62. T. Yamashita and M. A. Vannice, *J. Catal.* 163 (1996) 158.
63. S.-J. Huang, A. B. Walters and M. A. Vannice, *Appl. Catal. B* 17 (1998) 183.
64. S.-J. Huang, A. B. Walters and M. A. Vannice, *J. Catal.* 173 (1998) 229.

65. X. Zhang, A. B. Walters and M. A. Vannice, *Catal. Today* 27 (1996) 41.
66. X. Zhang, A. B. Walters and M. A. Vannice, *J. Catal.* 155 (1995) 290.
67. D. H. Olson, W. O. Haag and R. M. Lago, *J. Catal.* 61 (1980) 390.
68. H. A. Benesi and B. H. C. Winquist, *Adv. Catal.* 27 (1978) 97.
69. R. J. Gorte, *Catal. Lett.* 62 (1999) 1.
70. "Characterization of Solid Acids", F. Schuth, K. Sing and J. Weitkamp, Eds., Wiley, 2002.
71. K. J. Yoon, P. W. Walker, Jr., L. N. Mulay and M. A. Vannice, *Ind. & Eng. Chem. Prod. Res. Devel.* 22 (1983) 519.
72. L. N. Mulay, A. V. Prasad Rao, P. L. Walker, Jr., K. J. Yoon and M. A. Vannice, *Carbon* 23 (1985) 501.
73. K. J. Yoon, "Benzene Hydrogenation over Supported Iron Catalysts", Ph.D. Thesis, Pennsylvania State University, 1982.

Problem 3.1

A boron-doped carbon was prepared by adding 0.1 wt % B to a graphitized carbon black (Monarch 700, Cabot Corp.) and then heat treating this material, designated BC-1, at 2773 K under Ar [71,72]. Its surface area was determined by measuring N_2 adsorption at 80 K and using the BET equation (Eq. 3.4). The equilibrium N_2 uptakes versus the pressure are listed in the table below. What is the surface area of this carbon? If the heat of liquefaction for N_2 is 1.34 kcal mole^{-1}, what is the estimated heat of adsorption in the first monolayer? The P_o value for N_2 at the actual temperature measured was $P_o = 732$ Torr (760 Torr = 1 atm).

N_2 physisorption on carbon (from ref. 73).	
P (Torr)	Uptake, n (μmole N_2 g^{-1})
26.8	1166
86.2	1286
140.7	1395
190.0	1511
232.1	1636
278.0	1793

Problem 3.2

In a study of iron catalysts, the BET surface areas of a number of materials were determined using N_2 or Ar physisorption near 80 K [71–73]. The P_o values vary slightly because the bath temperature varied slightly around 77–80 K. What were the surface areas of the following solids based on the data provided below? What were the C values and the heats of adsorption for the initial monolayer of adsorbate? The heat of liquefaction for N_2 is 1.34 kcal mole^{-1}.

a) Silica (SiO_2) (Cab-O-Sil, Grade 5, Cabot Corp) – N_2 adsorption, assume $P_o = 725$ Torr (760 Torr = 1 atm). (Ref. 73)

Pressure (Torr)	Uptake (μmole N_2 g^{-1})
5.0	1067
27.2	1446
67.1	1763
123.0	1996
170.0	2142
207.1	2243
241.2	2359

b) Fresh reduced ferric oxide (Fe_2O_3) (Johnson Matthey, ultrapure) – N_2 adsorption, assume $P_o = 737.5$ Torr. (Ref. 73)

Pressure (Torr)	Uptake (μmole N_2 g^{-1})
61.1	4.29
128.0	8.04
185.5	11.34
238.7	14.17
297.8	17.53
350.2	20.79

c) Used reduced bulk iron (Johnson Matthey, ultrapure) – Ar adsorption (at 77 K), assume $P_o = 186$ Torr. Assume the cross-sectional area of an Ar atom is 13.9 Å [10], and the heat of liquefaction for Ar is 1.55 kcal mole^{-1}. (Ref. 73)

Pressure (Torr)	Uptake (μmole Ar g^{-1})
9.9	1.40
23.1	1.70
36.0	1.88
46.1	2.04
54.7	2.16
62.1	2.29

Problem 3.3

The following H_2 chemisorption results are reported for a 1.5% Rh/Al_2O_3 catalyst. What is the amount of hydrogen chemisorbed on Rh? What is the dispersion (fraction exposed) of the Rh?

H$_2$ pressure (Torr)	H$_2$ Uptake (μmole/g catalyst)
50	40.0
75	50.0
100	60.0
150	65.0
200	70.0
250	75.0
300	80.0
350	85.0
400	90.0

Problem 3.4

An O$_2$ adsorption isotherm was obtained at 443 K for a 2.43% Ag/SiO$_2$ catalyst, and the uptake results are given below. No irreversible adsorption occurs on the silica. What is the dispersion of the silver?

O$_2$ pressure (Torr)	Uptake (μmole O$_2$/g catalyst)
35	42.0
70	51.0
100	57.0
137	60.0
175	62.0
245	67.0
315	72.0

4
Acquisition and Evaluation of Reaction Rate Data

For purposes of kinetic modeling, it is important to collect reaction rate data that are free from experimental artifacts. Various types of reactors can be used to acquire these data, and the first portion of this chapter discusses these reactors. The second half of the chapter describes models which introduce the effect of mass and heat transfer gradients on the observed reaction rate, and it then provides different methods to evaluate the presence or absence of such artifacts in both gas-phase and liquid-phase reactions involving porous catalysts.

4.1   Types of Reactors

Different types of reactors can be utilized to conduct kinetic runs and to obtain reaction rate data as different reaction parameters are changed, such as temperature, concentration or partial pressure, catalyst loading, metal dispersion and so forth. Although batch and semi-batch reactors are frequently used in the chemical and pharmaceutical industries to manufacture limited quantities of a material, which is usually an expensive specialty product, they are not necessarily the reactor of choice in the laboratory to study heterogeneous catalysts. When investigating solid catalysts it is much more common to utilize flow systems, such as a plug-flow reactor (PFR) or a continuous-flow stirred tank reactor (CSTR). A short discussion of each type of reactor should be beneficial, and details of the derivations of the design equation for each can be found in numerous texts on reactor design. The following references represent but a few of the many books dedicated to this latter topic [1–5].

4.1.1   Batch Reactor

A batch reactor represents a closed system, i.e., no material crosses its boundaries, and the design equation is obtained by a mass balance on one of the species involved in the reaction, which is presumed to be the limiting reactant unless otherwise stated. For a well-stirred reactor with a uniform composition throughout its volume, the rate of accumulation of this species (designated A) is equal to the rate of its disappearance by chemical reaction:

$$[\text{Rate of accumulation of A}] = -\left[\begin{array}{l}\text{Rate of disappearance of A}\\ \text{by chemical reaction}\end{array}\right] \quad (4.1)$$

The rate of accumulation is expressed from equation 2.3 as follows, where N_{A_o} is the number of moles of A present at zero fractional conversion $(f_A = 0)$:

$$dN_A/dt = \nu_A \, d\xi/dt = -N_{A_o} \, df_A/dt \quad (4.2)$$

where the fractional conversion, f_A, is defined as:

$$f_A = (N_{A_o} - N_A)/N_{A_o} = 1 - (N_A/N_{A_o}) \quad (4.3)$$

The rate of disappearance due to reaction in the reactor volume, V_r, actually occupied by the reacting fluid is $(-r_A)V_r$, where r_A is the rate per unit volume.
Thus,

$$N_{A_o} \, df_A/dt = (-r_A)V_r \quad (4.4)$$

and the design equation is obtained by rearranging and integrating, i.e.,

$$t_2 - t_1 = N_{A_o} \int_{f_{A_1}}^{f_{A_2}} \frac{df_A}{(-r_A)V_r} \quad (4.5)$$

and if the initial conversion $f_{A_1} = 0$ at time $t = 0$, which is the typical situation, then:

$$t = N_{A_o} \int_0^{f_A} \frac{df_A}{(-r_A)V_r} \quad (4.6)$$

In constant-volume batch reactors, equation 4.6 can also be written as

$$t = C_{A_o} \int_0^{f_A} \frac{df_A}{(-r_A)} = -\int_{C_{A_o}}^{C_A} \frac{dC_A}{(-r_A)} \quad (4.7)$$

where C_A is the concentration of species A. If the reactor volume is not constant, but it varies linearly with fractional conversion:

$$V_r = V_{r_o}(1 + \delta_A f_A) \quad (4.8)$$

$$\text{where} \quad \delta_A = \frac{V(\text{at } f_A = 1) - V(\text{at } f_A = 0)}{V(\text{at } f_A = 0)} \quad (4.9)$$

then equation 4.6 can accommodate this change in the following way:

$$t = N_{A_o} \int_0^{f_A} \frac{df_A}{(-r_A)V_{r_o}(1 + \delta_A f_A)} = C_{A_o} \int_0^{f_A} \frac{df_A}{(-r_A)(1 + \delta_A f_A)} \qquad (4.10)$$

Note that the rate term $(-r_A)$ will be a positive value because the stoichiometric coefficient ν_A in the rate expression is negative (see chapter 2.1). Obviously in such reactors, sufficient mixing must be provided to assure a homogeneous system exists with no heat or mass transfer limitations.

Illustration 4.1 – Kinetic Behavior in a Batch Reactor

The gas-phase decomposition of nitrous oxide obeys second-order kinetics over a porous oxide catalyst at 573K, and it is irreversible, thus it can be written:

(1) $$2N_2O \overset{k}{\Longrightarrow} 2N_2 + O_2$$

Starting with pure N_2O at 1.0 atm in a well stirred, isothermal, constant-volume batch reactor containing 1.0 g of the catalyst, which has a density of $1.0\,g\,cm^{-3}$, after 10.0 s the fractional conversion is 0.093. What is the rate constant, k? What conversions are obtained after 1.0 s and 10.0 min? If a constant-pressure batch reactor is used instead with the same amount of catalyst, what times are required to achieve the same three fractional conversions?

Solution

The rate can be defined in terms of the disappearance of the reactant, N_2O, which has a stoichiometric number of -2 in reaction (1); therefore, as stated in Chapter 2.3:

$$r_m = \frac{1}{m} d\xi/dt = \left(\frac{1}{\nu_{N_2O}m}\right) dN_{N_2O}/dt = kC_{N_2O}^2 \qquad (1)$$

thus the rate for N_2O is:

$$-r_{N_2O} = -dN_{N_2O}/dt = 2\,mkC_{N_2O}^2 \qquad (2)$$

If the conversion of N_2O, f, is zero at $t = 0$, then equation 4.7 for a constant-volume batch reactor is applicable:

$$t = C_{N_2O_o} \int_0^f df/(-r_{N_2O}) = C_{N_2O_o} \int_0^f df/2\,mkC_{N_2O}^2 \qquad (3)$$

and from a rearrangement of equation 4.3 one has

$$N_{N_2O} = (1 - f)N_{N_2O_o} \qquad (4)$$

which leads to

$$C_{N_2O} = (1 - f)C_{N_2O_o} \quad \text{(for constant volume systems)} \qquad (5)$$

Substituting equation 5 into equation 3 and integrating gives

$$f/(1 - f) = 2\,mkC_{N_2O_o}t \tag{6}$$

For an ideal gas the initial concentration of N_2O is:

$$C_{N_2O_o} = N_{N_2O_o}/V = P_{N_2O_o}/RT = 1\,atm \Big/ \left(\frac{82.06\,atm \cdot cm^3}{g\,mole \cdot K}\right)(573\,K)$$

$$= 2.13 \times 10^{-5}\,g\,mole\,cm^{-3} \tag{7}$$

and rearrangement of equation 6 gives:

$$k = (0.093/0.907)/2(1\,g\,cat)(2.13 \times 10^{-5}\,g\,mole\,cm^{-3})(10.0\,s)$$

$$= 241\,cm^6\,g\,mole^{-1}\,s^{-1}\,g^{-1}$$

Consequently, to calculate the fractional conversions obtained for the other two reaction times, one has:

$$f = t \Big/ \left(t + \frac{1}{2mkC_{N_2O_o}}\right) \tag{8}$$

where the value of $1/(2\,mkC_{N_2O_o})$ is 97.4 s, and the values are $f = 0.0102$ at $t = 1.0\,s$ and $f = 0.860$ at $t = 600\,s$.

If one now examines the problem using a reactor with a variable volume (constant P), equations 4.8–4.10 must be used. With the definition of δ provided by equation 4.9:

$$\delta = (3 - 2)/2 = 1/2 \tag{9}$$

and

$$V_r = V_{r_o}(1 + f/2) \tag{10}$$

then from this equation and equation 4, the concentration at any time (or conversion) is:

$$C_{N_2O} = N_{N_2O}/V = N_{N_2O_o}(1 - f)/V_{r_o}(1 + f/2)$$

$$= C_{N_2O_o}(1 - f)/(1 + f/2) \tag{11}$$

Substituting equation 11 into equation 3 results in:

$$t = \left(\frac{1}{2\,mkC_{N_2O_o}}\right) \int_0^f \frac{(1 + f/2)df}{(1 - f)^2} \tag{12}$$

and integrating this gives:

$$t = \left(\frac{1}{2\,mkC_{N_2O_o}}\right) \left[\frac{1 + f/2}{1 - f} + 1/2\ln(1 - f)\right]_0^f \tag{13}$$

Using this last relationship, the times required for the three fractional conversions are:

t = 1.01 s for f = 0.010, t = 10.3 s for f = 0.093, and t = 807 s for f = 0.860. As expected, the reaction is slower because the increasing volume decreases the N_2O concentration more rapidly than that due only to reaction.

4.1.2   Semi-Batch Reactor

A semi-batch reactor is more difficult to analyze mathematically because at least one of the reactant or product species enters or leaves the system boundaries, thus specific applications should be modeled [1,5]. However, the most typical application for a semi-batch reactor is the presence of one reactant initially contained in a stirred tank reactor and a second reactant continuously added to the reactor, with no flow out of the reactor. The addition of a gas to participate in a liquid-phase reaction is one of the more common situations involving a semi-batch reactor, especially because the rate of addition of the gas can be controlled to keep its partial pressure essentially constant as well as providing quantitative information about the rate of reaction. In addition, there is frequently little or no change in the volume of the liquid phase. Well-mixed autoclave reactors coupled with gas pressure controllers, mass flow meters and computers can nicely provide continuous, real-time rate data related to heterogeneous catalysts used in such gas/liquid systems [6–8]. Again, it must be emphasized that experiments must be performed and/or calculations made to verify that no heat or mass transfer limitations exist.

4.1.3   Plug-Flow Reactor (PFR)

A plug-flow reactor (PFR), also known as a tubular reactor, is an open system with material entering and leaving its system boundaries, and it is invariably operated under steady-state conditions, thus the accumulation term for mass is zero. In modeling an ideal isothermal PFR to obtain a design equation, a mass balance is conducted on a tiny volume element (a plug) in a tubular reactor with the assumptions that: 1) radial mixing is infinitely rapid so that each plug of fluid is uniform in temperature, pressure, composition, etc.; 2) there is no longitudinal (axial) mixing between these plugs as they move through the reactor; and 3) all volume elements move through the reactor in the same amount of time. This model is depicted in Figure 4.1. The mass balance on limiting reactant A around this differential volume element, dV_r, is:

$$[\text{Flow rate of A in}] = [\text{Flow rate of A out}] + \begin{bmatrix} \text{Rate of disappearance of A} \\ \text{by chemical reaction} \end{bmatrix}$$

$$(4.11)$$

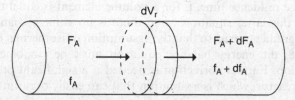

FIGURE 4.1. A differential volume element (dV_r) in a tubular (or plug flow) reactor with F and f being the flow rate and fractional conversion, respectively, of the limiting reactant A.

If F_A is the molal flow rate of A into this volume element, then equation 4.11 is described by:

$$F_A = (F_A + dF_A) + (-r_A)dV_r \qquad (4.12)$$

and thus

$$dF_A = r_A dV_r = -F_{A_o} df_A \qquad (4.13)$$

because

$$F_A = F_{A_o}(1 - f_A) \qquad (4.14)$$

Then, after rearrangement equation 4.13 becomes

$$dV_r/F_{A_o} = df_A/(-r_A) \qquad (4.15)$$

which can be integrated over the entire reaction volume to give

$$V_r/F_{A_o} = \int_{f_{A\ in}}^{f_{A\ out}} df_A/(-r_A) \qquad (4.16)$$

If the volumetric flow rate, V_o, is referenced to reactor inlet conditions, then

$$V_r/F_{A_o} = V_r/C_{Ao}V_o = \frac{\tau}{C_{Ao}} \qquad (4.17)$$

and this can be combined with equation 4.16 to give another useful form of the design equation for a PFR using space time, τ, i.e.:

$$\tau = V_r/V_o = C_{A_o} \int_{f_{A\ in}}^{f_{A\ out}} df_A/(-r_A) \qquad (4.18)$$

which, for constant volume systems only ($\delta = 0$), can also be written as:

$$\tau = -\int_{C_{A\ in}}^{C_{A\ out}} dC_A/(-r_A) \qquad (4.19)$$

Note that this equation with space time, τ, is analogous to equation 4.7 written for a batch reactor using real time, t.

The average residence time, $\bar{t}$, for a volume element is equal to the space time only in this latter situation when there is no volume change. Also, if temperature gradients exist so that the assumption of isothermal operation is inappropriate, the energy balance equation must be combined with the design equation. Further correction is needed if a significant pressure drop exists in the reactor, which is a situation that can easily occur in a fixed-bed reactor [1,4].

In such a fixed-bed reactor, once the catalyst packing density is known (mass catalyst/volume), equation 4.16 is easily modified to give:

$$W_c/F_{A_o} = \int_{f_{A\,in}}^{f_{A\,out}} df_A/(-r_{m_A}) \tag{4.20}$$

where W_c is the weight of catalyst in the reactor and $(-r_{m_A})$ is now the rate of reaction per unit mass of catalyst. In laboratory reactors, W_c is always known as is F_{A_o}, $f_{A\,in}$ is typically zero, and $f_{A\,out}$ is measured. It is the function $r_{m_A} = f(T, P_i)$, where P_i is the partial pressure of the components in the system, that is not known and must be determined by the experimenter. Even if the mathematical functionality of r_{m_A} is not known, its value can be determined by graphical integration of equation 4.20, i.e., by calculating the area under the curve of a plot of $\left(\frac{1}{-r_{m_A}}\right)$ vs. f_A when f_A is increased and the system is operated as an integral reactor [5].

Illustration 4.2 – Kinetic Behavior in a Plug Flow Reactor

What is the required flow rate of ortho-H_2 into a small fixed-bed PFR to achieve a conversion of 10.0% para-H_2 in the following system? The reversible reaction is:

$$(1) \qquad\qquad o - H_2 \underset{k_{-1}}{\overset{k_1}{\rightleftarrows}} p - H_2$$

The reactor contains 2.00 g of a Ni/Al_2O_3 catalyst, and it operates at 77.0 K and 40.0 psig (3.72 atm). The rate can be represented by a unimolecular Langmuir-Hinshelwood expression (See Chapter 7):

$$r_m = \frac{k_1[o - H_2] - k_{-1}[p - H_2]}{1 + K_{ad_o}[o - H_2] + K_{ad_p}[p - H_2]} \tag{1}$$

The forward rate constant is $k_1 = 1.10\,cm^3\,s^{-1}\,g\,cat^{-1}$, and the equilibrium constant for this reaction is $K_1 = 0.503$. The adsorption equilibrium constant is the same for both forms of H_2 and is $K_{ad} = 1.06 \times 10^3\,cm^3\,g\,mole^{-1}$ [2]. Thus the rate can be rewritten as:

$$r_m = (k_1/(1 + K_{ad}[H_2]))\,([o - H_2] - [p - H_2]/K_1)$$
$$= k[o - H_2] - k'[p - H_2] \tag{2}$$

Solution

The total concentration of H_2 is equal to the initial concentration of $o - H_2$, which is

$$[H_2] = [o - H_2]_o = \frac{N_{H_2}}{V} = P/RT = (3.72 \, \text{atm})/\left(\frac{82.06 \, \text{atm} \cdot \text{cm}^3}{\text{g mole} \cdot \text{K}}\right)(77 \, \text{K})$$

$$= 5.89 \times 10^{-4} \, \text{g mol cm}^{-3} \tag{3}$$

The equilibrium conversion must be determined, and if f is the fractional conversion of $o - H_2$ in a closed system, then with $[o - H_2]_o$ equal to 1 mole, $K_1 = [p - H_2]/[o - H_2] = f/(1 - f) = 0.503$ and f at equilibrium is 0.335, consequently, the reverse reaction is significant at a fractional conversion of 0.10.

The applicable design equation for such a reactor containing a solid catalyst is equation 4.20:

$$W_c/F = \int_0^f df/(-r_m) \tag{4}$$

where W_c is the weight of the catalyst and F is the molar flow rate of $o - H_2$.

In equation 2, the constant k is equal to:

$$k = \frac{k_1}{(1 + K_{ad}[H_2])} = \frac{1.1 \, \text{cm}^3 \, \text{s}^{-1} \, \text{g cat}^{-1}}{1 + (1.06 \times 10^3 \, \text{cm}^3 \, \text{g mole}^{-1})(5.89 \times 10^{-4} \, \text{gmole cm}^{-3})}$$

$$= 0.677 \, \text{cm}^3 \, \text{s}^{-1} \, \text{g cat}^{-1} \tag{5}$$

and the constant k' is equal to:

$$k' = k/K_1 = 0.677/0.503 = 1.346 \, \text{cm}^3 \, \text{s}^{-1} \, \text{g cat}^{-1} \tag{6}$$

The substitution of equation 2 into equation 4 gives:

$$W_c/F = \int_0^{0.10} df/(k[o - H_2] - k'[P - H_2])$$

$$= \int_0^{0.10} df/(k[o - H_2]_o(1 - f) - k'[o - H_2]_o f)$$

$$= \frac{1}{[o - H_2]_o} \int_0^{0.10} \frac{df}{k - (k + k')f} \tag{7}$$

Integration of equation 7 gives:

$$W_c/F = \frac{1}{[o - H_2]_o} \left(\frac{-1}{k + k'}\right) [\ln(k - (k + k')f)]_0^{0.10}$$

$$= \frac{1}{(o - H_2)_o} \left(\frac{-1}{2.022}\right) (\ln(0.677 - 0.202) - \ln 0.677)$$

$$= 0.175/[o - H_2]_o \tag{8}$$

thus

$$F = W_c[o - H_2]_o/0.175$$
$$= (2\,g\,cat)(5.89 \times 10^{-4}\,g\,mole\,cm^{-3})/(0.175\,cm^{-3}\,s\,g\,cat)$$
$$= 6.73 \times 10^{-3}\,g\,mole\,o - H_2\,s^{-1}$$

The operation of a PFR in a differential mode, i.e., at low values of Δf_A such that $df_A \cong \Delta f_A$, provides some real benefits in regard to acquiring experimental data. Use of a differential reactor of any type helps to eliminate any heat and mass transfer limitations, although tests should still be conducted to verify their absence but, perhaps more importantly, it simplifies the rate expression, r_{m_A}, as this function is now approximately constant throughout the reactor whether a change in the volume of the system occurs or not, thus, if $f_{A\,in} = 0$ and $f_{A\,out} = f_A$, equation 4.20 is simply

$$W_c/F_{A_o} = \left(\frac{1}{-r_{m_A}}\right)_{av} f_A \qquad (4.21)$$

where typically $f_A \leq 0.1$. The average rate in the catalyst bed

$$(-r_{m_A})_{av}\,(mole\,time^{-1}\,mass\,catalyst^{-1}) = (F_{A_o}/W_c)f_A \qquad (4.22)$$

is now determined at average concentrations or partial pressures that are known within specified limits which are determined by $f_{A\,out}$, i.e.,

$$N_{A_{av}} = N_{Ao}(1 - f_A/2) \pm f_A/2$$
$$N_{i_{av}} = N_{i_o} + v_i/v_A[N_{A_o}(1 - f_A/2) - N_{A_o}] = N_{i_o} - \frac{v_i N_{A_o} f}{2v_A} \qquad (4.23)$$

because from equation 2.1

$$\xi = \frac{N_i - N_{i_o}}{v_i} \qquad (4.25)$$

and

$$N_i = N_{i_o} + v_i/v_A(N_A - N_{A_o}) \qquad (4.26)$$

Therefore, average concentrations, $C_{i_{av}}$, are

$$C_{i_{av}} = \frac{N_{i_{av}}}{V} = \left[\left(N_{i_o} - \frac{v_i N_{A_o} f_A/2}{2v_A}\right) \pm f_A/2\right]\Big/V \qquad (4.27)$$

and the average partial pressures are:

$$P_{i_{av}} = C_{i_{av}}RT \qquad (4.28)$$

Consequently, if f_A is quite low, say 0.05, then the error in rate associated with the limiting reactant is only $\pm 2.5\%$ while the quantities of products are very low and can usually be ignored, if so desired. Rate dependencies on product species are typically obtained by adding known amounts of these compounds to the feed stream. By conducting isothermal runs at different values of P_i, a rate function in the form of the law of mass action:

$$r_{m_A} = k\Pi_i P_i^{a_i} = k'\Pi_i C_i^{b_i} \qquad (4.29)$$

can be determined, and this type of rate expression is typically referred to as a power rate law. Obtaining this functionality at various temperatures with $\Pi_i P_i^{a_i}$ held constant allows the temperature dependence of the rate constant, with E representing the activation energy,

$$k = Ae^{-E/RT} \qquad (4.30)$$

to be calculated via an Arrhenius plot.

Illustration 4.3 – Kinetic Behavior in a Differential PFR

The reaction between sulfur vapor and methane was conducted in a small silica tube reactor (a PFR) containing 30.0 g of an oxide catalyst [2]. In one run at 673 K and a pressure of 1.0 atmosphere, the quantity of carbon disulfide produced in a 10.0-minute run was 0.100 g. Assume that all the sulfur present is the molecular species S_2 so that the stoichiometry is:

(1) $$2\,S_2 + CH_4 \Longrightarrow CS_2 + 2\,H_2S$$

The flow rate of sulfur vapor into the reactor was 0.238 g mole S_2 per hour in this steady-state run, and the flow rate of methane was 0.119 g mole CH_4 per hour. There was no H_2S or CS_2 in the feed stream. The rate equation at this temperature is:

$$r_m = \frac{1}{m} dN_{CS_2}/dt = kP_{CH_4}P_{S_2} \qquad (1)$$

where P is the partial pressure in atm.

Assume the system behaves as a differential reactor and calculate the rate constant k in units of g mole h^{-1} g cat^{-1} atm^{-2}.

Solution

First, let us determine the fractional conversion to see how valid the assumption of a differential reactor is. The molar flow rate of CS_2 produced is:

$$\left(\frac{0.100\,\text{g CS}_2}{10.0\,\text{min}}\right)\left(\frac{60\,\text{min}}{h}\right)\left(\frac{\text{g mole CS}_2}{76.0\,\text{g CS}_2}\right) = 0.00789\,\text{g mole CS}_2\,h^{-1} \qquad (2)$$

The fractional conversion is:

$$\left(\frac{2\,\text{g mole S}_2\,\text{reacted}}{\text{g mole CS}_2\,\text{formed}}\right)\left(\frac{0.00789\,\text{g mole CS}_2}{h}\right)\bigg/\left(\frac{0.238\,\text{g mole S}_2\,\text{fed}}{h}\right)$$

$$= 0.0663 \qquad (3)$$

and the rate of reaction under these conditions is:

$$r_m = \left(\frac{0.00789 \text{ g mole } CS_2}{h}\right) \bigg/ (30.0 \text{ g cat}) \qquad (4)$$

$$= 2.63 \times 10^{-4} \text{ g mole } CS_2 \text{ h}^{-1} \text{ g cat}^{-1}$$

Because of equation 2.3

$$d\xi/dt = \frac{1}{\nu_i} dN_i/dt = -1/2 \, dN_{S_2}/dt = dN_{CS_2}/dt \qquad (5)$$

and

$$-dN_{S_2}/dt = (-r_{S_2}) = 2 \, dN_{CS_2}/dt = 2 \, r_{CS_2} \qquad (6)$$

The design equation for a fixed-bed PFR is provided by equation 4.20, and if a differential reactor is assumed, then equation 4.21 is used:

$$W_c/F_{S_{2o}} = f/(-r_{S_2})_{av} = f/(k_{S_2} P_{CH_4} P_{S_2})_{av} \qquad (7)$$

The rate is therefore assumed constant using the average partial pressures for CH_4 and CS_2. For an ideal gas mixture, the partial pressure of component i is:

$$P_i = y_i P_t = \frac{N_i}{N_t} P_t = \frac{\dot{N}_i}{\dot{N}_t} P_t \qquad (8)$$

where y_i is the mole fraction of species i and P_t is the total pressure. The average composition of CS_2 in the reactor, expressed using a molar flow rate, $\dot{N}$, is:

$$(0 + 0.00789)/2 = 0.00395 \text{ g mole } CS_2 \text{ h}^{-1}$$

This is a constant volume system, $\sum_i \nu_i = 0$, so mass balances based on the reaction stoichiometry give average flow rates of:

$$F_{CS_2} = 0.00395$$
$$F_{H_2S} = 0.00790$$
$$F_{CH_4} = 0.119 - 0.004 = 0.115$$
$$F_{S_2} = 0.238 - 2(0.004) = 0.230$$
$$\overline{F_t = 0.357 \text{ g mole h}^{-1}}$$

Thus the respective average partial pressures of CH_4 and S_2 are:

$$P_{CH_4} = (0.115 \text{ g mole h}^{-1})(1.0 \text{ atm})/(0.357 \text{ g mole h}^{-1}) = 0.322 \text{ atm}$$

and

$$P_{S_2} = (0.230)(1.0)/(0.357) = 0.644 \text{ atm}$$

Rearranging equation 7 gives:

$$k_{S_2} = f\, F_{S_{2o}}/W_c(P_{CH_4}P_{S_2})_{av} = \frac{(0.0663)(0.238\text{ gmoleS}_2\text{ h}^{-1})}{(30.0\text{ g cat})(0.322\text{ atm})(0.644\text{ atm})}$$

$$= \frac{2.54 \times 10^{-3}\text{ g mole S}_2}{\text{h g cat atm}^2} \tag{9}$$

or, as shown in equation 6

$$k_{CS_2} = k_{S_2}/2 = 1.27 \times 10^{-3}\text{ g mole CS}_2\text{ h}^{-1}\text{ g cat}^{-1}\text{ atm}^{-2}$$

If differential behavior is not assumed, then equation 4.20 gives:

$$W_c/F_{S_{2o}} = \int_0^f \frac{df}{(-r_{S_2})} = \int_0^{0.0663} \frac{df}{k_{S_2}P_{CH_4}P_{S_2}} \tag{10}$$

Mass balances on the four components are now:

$$F_{CS_2} = 0.119\,f$$
$$F_{H_2S} = 0.0238\,f$$
$$F_{CH_4} = 0.119(1 - f)$$
$$F_{S_2} = 0.238(1 - f)$$
$$\overline{F_t = 0.357\text{ g mole h}^{-1}}$$

with

$$P_{CH4} = 0.119(1 - f)(1)/0.357\text{ atm}$$

and

$$P_{S_2} = 0.238(1 - f)(1)/0.357$$

Substituting these latter two relationships into equation (10) gives:

$$W_c/F_{S_{2o}} = (0.357)^2/(0.119)(0.238)\,k_{S_2}\text{ atm}^2$$

$$\int_0^{0.0663} \frac{df}{(1 - f)^2} = (4.50/k_{S_2})\frac{f}{1 - f}\text{ atm}^{-2} \tag{11}$$

Therefore, k_{S_2} is

$$k_{S_2} = \frac{4.50(0.238\text{ g mole S}_2\text{ h}^{-1})}{30.0\text{ g cat}\left(\dfrac{0.0663}{1-0.0663}\right)\text{ atm}^{-2}} = 2.53 \times 10^{-3}\text{ g mole S}_2\text{ h}^{-1}\text{ g cat}^{-1}\text{ atm}^{-2}$$

$$\tag{12}$$

The difference between the values for k_{S_2} given by equations 9 and 12 is clearly very minor at this low conversion.

4.1.4 Continuous Flow Stirred-Tank Reactor (CSTR)

This type of reactor, which is also known as a continuous stirred tank reactor, a stirred-flow reactor or a back-mix reactor, has the advantage of analytical simplicity because perfect mixing in the reactor is assumed; consequently, system properties are uniform throughout the reactor and the composition of the exit stream is the same as that in the reactor. Thus at steady state a mass balance around the reactor on the limiting reactant, A, is the same as equation 4.11, but in terms of the symbols used in section 4.1.3, this becomes simply an algebraic expression:

$$F_{A_{in}} = F_{A_{out}} + (-r_{A_F})V_r \qquad (4.31)$$

where V_r again represents the volume actually occupied by the reacting mixture and the rate $(-r_{A_F})$ is that evaluated at outlet conditions. Rewriting equation 4.31 in terms of fractional conversion gives the straightforward relationship:

$$(-r_{A_F}) = F_{A_o}(f_{A_{out}} - f_{A_{in}})/V_r \qquad (4.32)$$

which, if $f_{A_{in}} = 0$ and $f_{A_{out}} = f_A$ is just:

$$(-r_{A_F}) = F_{A_o}f_A/V_r \qquad (4.33)$$

It is clear that in a CSTR the lowest possible reaction rate occurs, assuming the reactions have positive reaction orders. Equations 4.32 and 4.33 are completely general and independent of any changes in density (whether $\delta_A = 0$ or not) or differences between inlet and reactor temperatures. The geometry of the reactor is not important, but the effective volume, V_r, must be known. With V_o and C_{A_o} again representing the volumetric flow rate and concentration of A, respectively, at reactor inlet conditions and zero conversion, because $\tau = V_r/V_o$, equation 4.33 can be rewritten as:

$$(-r_{A_F}) = C_{A_o}f_A/\tau \qquad (4.34)$$

In the case of constant density systems, such as liquid-phase reactions, equation 4.34 can also be expressed as:

$$(-r_{A_F}) = (C_{A_o} - C_{A_{out}})/\tau \qquad (4.35)$$

and CSTRs have been used in the laboratory for decades to study such reactions. Specially designed CSTRs have subsequently been developed to allow their application to gaseous reaction systems catalyzed by solids so that kinetics can be determined for heterogeneous catalytic systems [9–11]. Finally, it is worth noting that CSTR performance can be closely approximated with a PFR if a high recycle ratio is employed with the latter reactor [1–5]. Again, as with a PFR, dependencies of the rate on temperature and concentration or partial pressures can be determined with a CSTR to get a rate expression in the form of a power rate law applicable over a specified range of reaction conditions.

4.2 Heat and Mass Transfer Effects

In commercial reactors, heat and mass transfer effects frequently impact upon the overall performance of the reactor because rates are maintained as high as possible to maximize yields. However, concentration and temperature gradients can be built into the overall reactor design models provided accurate kinetic rate expressions are available. The opposite is not true, though, and accurate kinetic rate equations can seldom, if ever, be extracted from data obtained under the influence of significant heat and/or mass transport limitations. Thus it is important that the experimenter conduct kinetic runs under reaction conditions that guarantee the rate data are acquired in the regime of kinetic control. To get a perspective on the physical situation, let us examine a simple model of a porous catalyst particle, which might represent one of those comprising the granule in Figure 3.1, as illustrated in Figure 4.2 for a gas-phase exothermic reaction. This subject has been discussed in a number of texts [2,5,12,13, for example], but the following development is drawn heavily from that used by Carberry [12].

Before a reactant molecule can react, it must be transported from the well-mixed, homogeneous bulk phase to the surface of the catalyst particle and

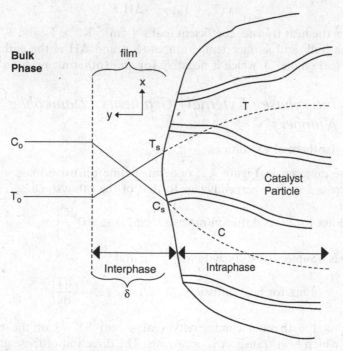

FIGURE 4.2. A diagram depicting concentration and temperature changes from the bulk phase through a stagnant film of thickness δ to a particle surface and then through the porous catalyst particle, assuming an exothermic reaction occurs.

then, for porous materials containing active sites distributed within their structure, it must further diffuse into the pores. Interphase gradients can exist between the bulk and solid phases, i.e., diffusive-convective transport processes link the source of reactants (bulk phase) to the sink of reaction (catalyst particle surface). Consequently, here diffusion and convection occur **in series** with the reaction. Intraphase gradients are confined within the particle to the local reaction zone; therefore, pore diffusion occurs **simultaneously** with the reaction.

At steady-state conditions, a mass balance shows the rate of transport of the reactants through a thin film surrounding the particle will be equal to the rate of reaction, i.e.,

$$k_g a(C_o - C_s) = \Re = kC_s^n \qquad (4.36)$$

where k_g is a gas-phase mass transfer coefficient (cm s^{-1}), a is the external interface area per unit volume (cm^{-1}), C_o and C_s are the bulk and surface concentrations (mole cm^{-3}), respectively, $\Re$ is the global (or effective) reaction rate (mole s^{-1} cm^{-3}), k is a reaction rate constant (s^{-1}(cm^3 mole^{-1})$^{n-1}$), and n is the reaction order. Typical units have been listed in parentheses with each parameter. In similar fashion, an energy balance on the film under steady-state conditions for heat transfer gives:

$$ha\,(T_s - T_o) = -\Delta H\,\Re \qquad (4.37)$$

where h is the heat transfer coefficient (cal s^{-1} cm^{-2} K^{-1}), T_o and T_s are the respective bulk and surface temperatures (K), and ΔH is the enthalpy of reaction (cal gmole^{-1}), which is negative for an exothermic reaction.

4.2.1 Interphase (External) Gradients (Damköhler Number)

4.2.1.1. Isothermal Conditions

From the diagram in Figure 4.2, one can define diffusive mass and heat fluxes across a surface perpendicular to that of the catalyst, i.e.,

$$\text{Flux for mass diffusion(moles s}^{-1}\text{cm}^{-2}) \equiv -D\frac{dC}{dy}\bigg|_{y=\delta} \qquad (4.38)$$

where D is a molecular diffusivity (cm^{2}s^{-1}), and

$$\text{Flux for heat diffusion(cal s}^{-1}\text{cm}^{-2}) \equiv -\lambda\frac{dT}{dy}\bigg|_{y=\delta} \qquad (4.39)$$

where λ is the thermal conductivity (cal s^{-1}cm^{-1} K^{-1}) of the material through which heat transport is occurring. The direction of these gradients is indicated by y in Figure 4.2, and the fluxes are evaluated across the stagnant film thickness, δ.

First let us examine mass transport through this film under isothermal conditions by employing the continuity equations for mass (a mass balance) and for momentum (an energy balance). In this stagnant film, which can correspond to the laminar boundary layer that develops when a fluid passes over a flat surface, there is no motion of the fluid, hence the latter equation is irrelevant. The continuity equation for mass describes the spacial dependence of concentration in terms of the velocities parallel, u, and perpendicular, v, to the surface:

$$u\frac{\partial C}{\partial x} + v\frac{\partial C}{\partial y} = D\left(\frac{\partial^2 C}{\partial x^2} + \frac{\partial^2 C}{\partial y^2}\right) \tag{4.40}$$

(axial   (perpendicular (axial   (perpendicular
convection) convection)   diffusion)   diffusion)

Here the contribution of each term is also identified. As there is no fluid motion in this boundary layer (or film), the convection terms are zero, and if axial diffusion parallel to the surface is minimal and ignored, then equation 4.40 becomes simply:

$$\partial^2 C/\partial y^2 = 0 \tag{4.41}$$

with boundary conditions of $C = C_o$ at $y = \delta$ and $C = C_s$ at $y = 0$. Integrating this gives:

$$C = (C_o - C_s)(y/\delta) + C_s \tag{4.42}$$

thus the diffusive flux is

$$D\frac{dC}{dy}\bigg|_{y=\delta} = D\left(\frac{C_o - C_s}{\delta}\right) = k_g(C_o - C_s) \tag{4.43}$$

and so

$$k_g = D/\delta \tag{4.44}$$

One can now examine the influence of interphase mass transfer limitations on heterogeneous reaction rates. For simplicity let us assume isothermal, steady-state conditions with the reaction being constrained to just the surface of the particles, as would be the case for a nonporous material, and now equation 4.36 describes the situation. Keep in mind that $\Re$ is the effective (or global), experimentally measurable reaction rate. Consider a 1st-order reaction ($n = 1$) where the unknown (or unobservable) concentration C_s is now:

$$C_s = \frac{C_o}{1 + \frac{k}{k_g a}} = \frac{C_o}{1 + Da_o} \tag{4.45}$$

where Da is a dimensionless Damköhler number representing the ratio of the chemical reaction rate to the bulk mass transport rate, and the subscript zero refers to bulk conditions. An isothermal interphase effectiveness factor can now be defined to evaluate the influence of external mass transport on the global rate. Substituting equation 4.45 into 4.36 gives:

$$\Re = kC_s = \frac{kC_o}{1 + Da_o} \tag{4.46}$$

and if $\Re_o$ is the rate with no transport limitations (kinetic control regime and $C_o = C_s$) then the external effectiveness factor, $\bar{\eta}$, is:

$$\bar{\eta} = \Re/\Re_o = \frac{1}{1 + Da_o} \tag{4.47}$$

Thus, when diffusion is rapid compared to the reaction rate, $Da_o \to 0$, $C_s \cong C_o$ and $\Re = kC_o$; alternatively, if the surface reaction is very rapid relative to diffusion, $Da_o \to \infty$, $C_s \to 0$ and $\Re = k_g a C_o$. In the latter case, a very low, near-zero activation energy will prevail.

Consequently, in general

$$Da_o = \frac{kC_o^{n-1}}{k_g a} = \frac{\text{chemical reaction rate}}{\text{mass transfer rate}} \tag{4.48}$$

and no matter what the kinetic reaction order is (other than being positive), the global rate $\Re$ always becomes first order in the regime of external mass transfer control (large Da_o), as illustrated in Figure 4.3.

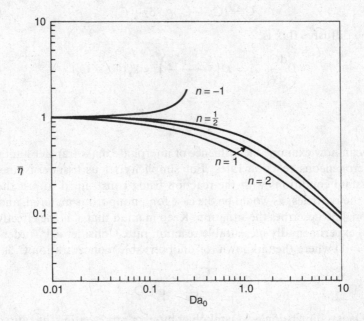

FIGURE 4.3. Isothermal external catalytic effectiveness for reaction order n. (Reprinted from ref. 12, copyright © 1976, with permission from Alison Carberry Kiene, executor)

If the rate constant k is known, then $\bar{\eta}$ can be evaluated because 'a' can be measured and k_g can be estimated to give a value for Da_o and then:

$$\Re = \bar{\eta}kC_o = \bar{k}f(C_o) \tag{4.49}$$

However, k is seldom known a priori and only $\bar{k}$ is measured, thus Da is an unobservable and cannot be calculated, but

$$\bar{\eta}Da_o = \left(\frac{\Re}{\Re_o}\right)\left(\frac{kC_o^n}{k_gaC_o}\right) = \frac{\Re}{k_gaC_o} \tag{4.50}$$

which is an observable as all quantities can either be measured or calculated. Only a relationship between $\bar{\eta}$ and $\bar{\eta}Da_o$ is needed, and this is provided by a heat balance. For a generalized, nonisothermal external effectiveness factor:

$$\bar{\eta} = \frac{k_sC_s^n}{k_oC_o} = \frac{k_s}{k_o}\left(\frac{C_s}{C_o}\right)^n \tag{4.51}$$

where the numerator is the rate at the surface with k_s varying with the surface temperature and the denominator is the rate with no gradients (bulk conditions). Equation 4.36 can be rearranged to give:

$$C_s/C_o = 1 - \frac{\Re}{k_gaC_o} = 1 - \bar{\eta}Da_o \tag{4.52}$$

and substituting this into equation 4.51 provides us with:

$$\bar{\eta} = \frac{k_s}{k_o}(1 - \bar{\eta}Da_o)^n \tag{4.53}$$

which is the relationship required. If thermal gradients are unimportant, $k_s = k_o$ and $\bar{\eta}$ can be determined readily using equation 4.53, i.e.,

$$\eta = (1 - \eta Da_o)^n \tag{4.54}$$

and this relationship is plotted in Figure 4.4 for different values of n.

4.2.1.2   Nonisothermal Conditions

For nonisothermal systems

$$\frac{k_s}{k_o} = e^{-E/R(\frac{1}{T_s} - \frac{1}{T_o})} = e^{-E/RT_o(1/t-1)} = e^{-\varepsilon_o(1/t-1)} \tag{4.55}$$

where $t = T_s/T_o$ and $\varepsilon_o = E/RT_o$. The steady-state heat balance given in equation 4.37 is rearranged and used to relate T_s and T_o, i.e.,

$$t = T_s/T_o = 1 - \frac{\Delta H\Re}{haT_o} = 1 - \left(\frac{\Delta Hk_gaC_o}{haT_o}\right)\left(\frac{\Re}{k_gaC_o}\right) = 1 + \bar{\beta} \cdot \bar{\eta}Da_o \tag{4.56}$$

where $\bar{\beta} = -\Delta Hk_gC_o/hT_o$ and both k_g and h can be calculated [12]. Finally, in terms of these definitions the nonisothermal effectiveness factor is:

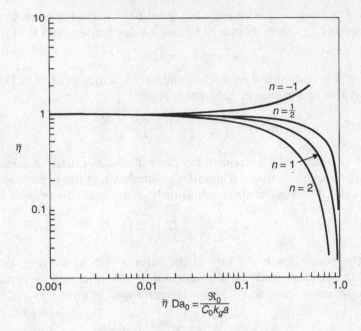

FIGURE 4.4. Isothermal external catalytic effectiveness in terms of observables for order n. (Reprinted from ref. 12, copyright © 1976, with permission from Alison Carberry Kiene, executor)

$$\bar{\eta} = (1 - \bar{\eta}Da_o)^n e^{-\varepsilon_o \left(\frac{-\beta \cdot \bar{\eta}Da_o}{1+\beta \cdot \bar{\eta}Da_o} \right)} \tag{4.57}$$

Thus with exothermic reactions this factor can significantly exceed unity if temperature gradients exist, as illustrated in Figure 4.5 for a 1st-order reaction.

4.2.2 Intraphase (Internal) Gradients (Thiele Modulus)

4.2.2.1. Isothermal Conditions

Let us now examine an isothermal reaction occurring simultaneously with diffusion inside the pore structure of a catalyst particle, which will be represented by a sphere of radius R_p, as shown in Figure 4.6. Other geometries can be chosen, but the forms of the final equation are all very similar and result in the same conclusions. At steady-state conditions, the material transported from the differential spherical volume between r and r + dr must equal that generated by chemical reaction, thus

$$d(\text{flux} \cdot \text{area}) = \Re dV \tag{4.58}$$

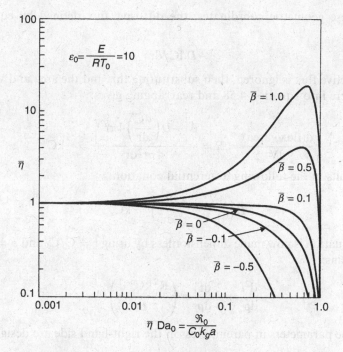

FIGURE 4.5. External nonisothermal effectiveness $\bar{\eta}$ versus observables (1st-order, $\varepsilon_o = 10$). (Reprinted from ref. 12, copyright © 1976, with permission from Alison Carberry Kiene, executor)

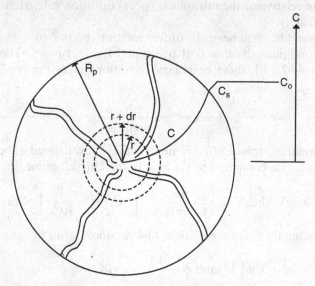

FIGURE 4.6. Schematic drawing of a spherical porous catalyst particle with radius R_p indicating concentration decreases going from the bulk (C_o) to the surface (C_s) and then through the pores (C).

With these geometric coordinates, the diffusive flux defined by equation 4.38 is

$$-DdC/dr \tag{4.59}$$

If convective flux is ignored, then substituting this and the area and volume of a sphere into equation 4.58 and rearranging gives:

$$\frac{d(\text{flux} \cdot \text{area})}{dV} = \Re = \frac{d\left[-D\left(\frac{dC}{dr}\right)4\pi r^2\right]}{4\pi r^2 dr} = -kC^n \tag{4.60}$$

This results in the following differential equation:

$$D\left(\frac{d^2C}{dr^2} + \frac{2}{r}\frac{dC}{dr}\right) = kC^n \tag{4.61}$$

If this equation is now made dimensionless by using $f = C/C_s$ and $\rho = r/R_p$, one obtains:

$$\frac{d^2f}{d\rho^2} + \frac{2}{\rho}\frac{df}{d\rho} = \left(\frac{R_p^2 k C_s^{n-1}}{D}\right)f^n \tag{4.62}$$

and if the parameters in parentheses on the right-hand side are designated:

$$R_p^2 k\, C_s^{n-1}/D = \phi^2 \tag{4.63}$$

a Thiele modulus, ϕ, is defined [14–16], which is a measure of the chemical reaction rate relative to the intraphase (pore) diffusion rate. Here D represents a diffusivity.

As an example, assume a 1st-order reaction occurs ($n = 1$) with the boundary conditions $df/d\rho = 0$ at $\rho = 0$ and $f = 1$ at $\rho = 1$. The solution to equation 4.62 with these boundary conditions gives the concentration profile:

$$f = C/C_s = \frac{\sinh \phi\rho}{\rho \sinh \phi} \tag{4.64}$$

If an isothermal intraphase effectiveness factor is now defined as before, i.e., the observable rate compared to the rate in the kinetic regime, then

$$\eta = \int_0^V kCdV/(kC_sV) = \frac{4\pi}{(4/3)\pi R^3 C_s}\int_0^R Cr^2 dr = \frac{1}{RC_s}\int_0^R Cdr \tag{4.65}$$

and substituting for C from equation 4.64 produces, after integration:

$$\eta = 3/\phi\left(1/\tanh \phi - \frac{1}{\phi}\right) = \frac{3}{\phi^2}(\phi \coth \phi - 1) \tag{4.66}$$

The same result is obtained if the solution is developed by viewing η as the ratio of the actual flux through the external surface of the spherical particle

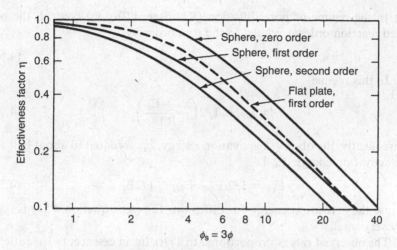

FIGURE 4.7. Effectiveness factors for power-law kinetics. For spheres, the abscissa is ϕ_s, while for a flat plate the abscissa is 3ϕ. (Reprinted from ref. 29, copyright © 1970, with permission from C. A. Satterfield)

to the rate (the flux) which would exist in the regime of kinetic control, i.e., no diffusional limitations, then

$$\eta = \frac{4\pi R_p^2 D (dC/dr)_{r=R_p}}{4/3\pi R_p^3 k_o C_s} = \frac{3D(dC/dr)_{r=R_p}}{R_p k_o C_s} \tag{4.67}$$

This gives the same equation for η when the derivative of $C/C_s = \left(\frac{\sinh \phi\rho}{\rho \sinh \phi}\right)$ is placed into equation 4.67. The dependence of η on ϕ is shown in Figure 4.7 for different reaction orders and also for a different geometry.

The influence of the internal effectiveness factor, η, on global rate thus has similarities to that of the external effectiveness factor, $\bar{\eta}$, in that: a) the higher the reaction order, the greater the diffusional effect; b) $\eta \to$ unity for small values of the Thiele modulus, ϕ, and similarly, $\bar{\eta} \to$ unity for small values of the Damköhler number, Da_o; and c) at large values of these two moduli, $\eta \cong 1/\phi$(for $\phi > 3$) and $\bar{\eta} \cong 1/Da_o$. Assuming that external mass transfer limitations have been removed ($C_s = C_o$), the effect of internal (pore) diffusion on the observed kinetics can be determined; i.e., for $\phi > 3$, $\eta \cong 1/\phi$ and

$$\Re = \eta k C_o^n = \frac{k C_o^n}{\phi} = \frac{k C_o}{R_p \left(\frac{k C_o^{n-1}}{D}\right)^{1/2}} = \frac{(kD)^{1/2}}{R_p} C_o^{(n+1)/2} = k_{ob} C_o^{n_{ob}} \tag{4.68}$$

with R_p representing a characteristic length, in this case the radius of the catalyst particle, and k_{ob} representing the observed apparent rate constant. Thus the following conclusions can be drawn:

a) In the regime of pore diffusion (Knudsen diffusion) control, the observed reaction order is related to the true order by

$$n_{ob} = (n + 1)/2 \qquad (4.69)$$

b) In this regime

$$\ln k_{ob} = 1/2 \left(\frac{E_t + E_D}{RT} \right) \qquad (4.70)$$

Consequently, the observed activation energy, E_{ob}, is equal to about half the true activation energy, E_t, i.e.,

$$E_{ob} = 1/2(E_t + E_D) \cong 1/2E_t \qquad (4.71)$$

because the activation energy for diffusion, E_D, (see equation 4.71) is typically very small.

c) The observed rate is proportional to $1/R_p$ or, in essence, to the ratio of the external surface to the volume (A/V). Thus one consequence of this analysis is the expectation that as a reaction moves from that of kinetic control to that of pore diffusion control as the temperature increases

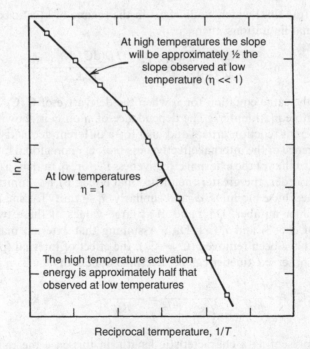

At high temperatures the slope will be approximately ½ the slope observed at low temperature ($\eta \ll 1$)

At low temperatures $\eta = 1$

The high temperature activation energy is approximately half that observed at low temperatures

Reciprocal termperature, $1/T$

FIGURE 4.8. Schematic representation of shift in activation energy when intraparticle mass transfer effects become significant. (Reprinted from ref. 1, copyright © 1977. This material is used by permission of John Wiley & Sons, Inc.)

($E_t \gg E_D$, thus k and hence ϕ increase significantly), the observed activation energy is decreased by about one-half, as demonstrated in Figure 4.8.

If necessary, an overall effectiveness factor incorporating both interphase and intraphase transport limitations can be determined [12]; however, it is very likely that either the kinetic control regime or one of the two transport control regimes will dominate, and this is discussed later.

4.2.2.2   Nonisothermal Conditions

Not a great deal will be said here regarding the derivation of nonisothermal effectiveness factors because it can be quite complicated, and details of a number of approaches can be readily obtained [5,12,17–19]. Two general statements can be made. First, for endothermic reactions, the effectiveness factor is decreased by both heat and mass transfer gradients. Second, for exothermic reactions, the effectiveness factor can be much greater than unity because of the Arrhenius factor, and numerous calculations have demonstrated this graphically [12]. One useful relationship for use in the laboratory is the following which, if obeyed, indicates that intraparticle heat transfer is rapid enough so that temperature gradients are insignificant [17]:

$$\frac{|\Delta H| \Re R_p^2}{\lambda T_s} < \frac{0.75 T_s R}{E_t} \tag{4.72}$$

where R is the gas constant, E_t is the true activation energy for the reaction, and the other symbols have been previously defined. If simultaneous heat and mass transfer effects are considered, it has been shown that for reaction orders of n = 1 or higher, $\eta = 1 \pm 0.05$ if

$$\frac{\Re R_p^2}{C_s D_{eff}} < \frac{1}{|n - \gamma \beta|} \tag{4.73}$$

where D_{eff} is the effective diffusivity in the pores of the catalyst, $\gamma = E_t / R T_s$ and $\beta = (-\Delta H D_{eff} C_s)/(\lambda T_s)$. In other words, the above criterion of isothermal operation is achieved when $|\gamma \beta| < 0.05\,n$.

4.2.2.3   Determining an Intraphase (Internal) Effectiveness Factor from a Thiele Modulus

Experimentally, how does one obtain the values needed to calculate an internal effectiveness factor? Clearly, if ϕ is known, then a correlation such as those shown in Figure 4.7 allows η to be determined directly; however, remember that in actuality

$$\phi = R_p \left(\frac{k C_s^{n-1}}{D_{eff}} \right)^{1/2} \tag{4.74}$$

where k and C_s are not observables, but the catalyst particle size can be measured and controlled, and the effective diffusivity, D_{eff}, in the Knudsen diffusion regime can be estimated once the average pore size is known. For a gaseous system, the bulk diffusivity D_b is:

$$D_b = \frac{\bar{v}\lambda_g}{3} \tag{4.75}$$

where λ_g is the mean free path (i.e., the distance between collisions) in the gas phase and $\bar{v}$ is the mean velocity:

$$\bar{v} = \left(\frac{8k_BT}{\pi m}\right)^{1/2} \tag{4.76}$$

where k_B is Boltzmann's constant and m is the mass of the molecular species. As the average pore diameter, d_p, becomes smaller than the mean free path, the diffusional process becomes more and more controlled by collisions between the molecular species and the pore walls, and the combined or effective diffusivity is now, depending on the reaction stoichiometry, approximately [2]:

$$D_{eff} = \frac{1}{1/D_b + 1/D_{Kn}} \tag{4.77}$$

where D_{Kn} is the Knudson diffusivity:

$$D_{Kn} = \bar{v}\frac{d_p}{3} \tag{4.78}$$

and in small pores where Knudsen diffusion dominates, $D_{Kn} \ll D_b$ and $D_{eff} \cong D_{Kn}$. Thus one approach to obtain η is an interactive method, that is, a reaction order n is selected, R_p is measured, D_{eff} is calculated, and $C_s = C_o$ is assumed, then a value for η is chosen and a value of ϕ is determined from equation 4.66 (or from $\eta = 3/\phi$ if ϕ is large). Equation 4.74 is equated to this value of ϕ and a value for the true rate constant, k, is calculated. To check this calculated value of k, remember that

$$\eta = \Re/\Re_o = \Re/kC_o^n = \eta kC_o^n/kC_o^n \tag{4.79}$$

consequently, one determines if the selected η value is obtained when k is substituted into this equation. If not, another iteration is made until agreement is attained.

Alternatively, from equation 4.68 one sees that for a 1st-order reaction, for example:

$$k = \frac{k_{ob}}{\eta} = \frac{\phi k_{ob}}{3\left[\dfrac{1}{\tanh \phi} - \dfrac{1}{\phi}\right]} \tag{4.80}$$

then, from the Thiele modulus:

$$k = \frac{\phi^2 D_{eff}}{R_p^2} \tag{4.81}$$

and equating equations 4.80 and 4.81 gives, after rearrangement:

$$\frac{k_{ob}R_p^2}{3D_{eff}} = \phi\left[\frac{1}{\tanh\phi} - \frac{1}{\phi}\right] \tag{4.82}$$

This can be solved directly for ϕ once the appropriate units for k are used so that all units cancel as equation 4.74 is dimensionless.

4.2.3 Intraphase Gradients (Weisz-Prater Criterion)

4.2.3.1 Gas-Phase or Vapor-Phase Reactions

Another approach to evaluate the influence of pore diffusion on a catalytic reaction is that taken by Weisz [20,21]. It is particularly useful because it provides a dimensionless number containing only observable parameters that can be readily measured or calculated. Let us again choose a spherical catalyst particle as our model, with volume V, surface area A, and radius R_p, as indicated in Figure 4.9. This figure also has axes to represent the decrease

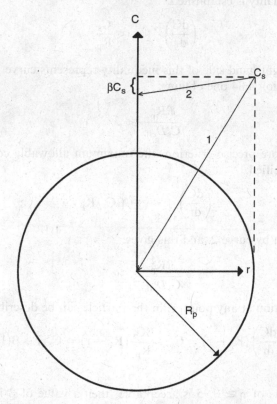

FIGURE 4.9. Depiction of concentration gradients for the Weisz-Prater criterion in a porous catalyst particle with radius R_p. The surface concentration is C_s. The maximum acceptable decrease in concentration within the pore is βC_s.

in concentration, C, from the surface concentration, C_s, as the reactant diffuses into the particle and reacts (curve 1). Under isothermal conditions at steady state, the rate of reaction of the limiting reactant within the particle must be equal to the diffusion rate across the surface of the particle into the pores, therefore

$$\Re V = A D_{eff} \left(\frac{dC}{dr} \right)_{r=R_p} \tag{4.83}$$

or, for a sphere:

$$\Re = \frac{3}{R_p} D_{eff} \left(\frac{dC}{dr} \right)_{r=R_p} \tag{4.84}$$

where $\Re$ is the observed rate and D_{eff} essentially represents Knudsen diffusivity in the pores (equation 4.78). As a general criterion, it is specified that a negligible decrease in the reaction rate requires that a negligible concentration gradient must exist, as indicated by curve 2 in Figure 4.9, thus the following inequality is established:

$$\left(\frac{dC}{dr} \right)_{r=R_p} \ll \frac{C_s}{R_p} \tag{4.85}$$

in which the right-hand side of this inequality represents curve 1. Substituting from equation 4.84 one obtains:

$$\frac{\Re R_p^2}{C_s D_{eff}} \ll 3 \tag{4.86}$$

Now, for a more precise criterion, the maximum allowable concentration gradient is specified:

$$\left(\frac{dC}{dr} \right)_{r=R_p} \leq \beta C_s / R_p \tag{4.87}$$

which is shown by curve 2, and this gives:

$$\frac{\Re R_p^2}{C_s D_{eff}} \leq 3\beta \tag{4.88}$$

The concentration at any point, r, in the particle can be described by:

$$C = C_s - \left(\frac{dC}{dr} \right)(R_p - r) = C_s - \frac{\beta C_s}{R_p}(R_p - r) = C_s[1 - \beta(1 - r/R_p)] \tag{4.89}$$

Now, if a value of $\eta \geq 0.95$ is acceptable, then a value of β is selected to obtain it. Again, if $\Re = kC^n$ and $\eta = \Re/\Re_{C_s}$ where $\Re_{C_s} = kC_s^n$ and:

$$\Re = \int_0^V kC^n dV = k \int_0^{R_p} C_s^n (4\pi r^2) dr \tag{4.90}$$

Substituting equation 4.89 into equation 4.90 gives:

$$\eta = \frac{3}{R_p^3} \int_0^{R_p} [1 - \beta(1 - r/R_p)]^n r^2 dr \tag{4.91}$$

Using a binomial expansion for small values of β [20], it can be shown that:

$$\eta = 1 - \frac{n\beta}{4} \tag{4.92}$$

If a second-order reaction is chosen as a reasonable upper limit, then with $n = 2$, $\beta = 0.95$, and the dimensionless Weisz criterion, N_{W-P}, frequently referred to as the Weisz-Prater number [21], is obtained for negligible diffusional limitations:

$$N_{W-P} = \frac{\Re R_p^2}{C_s D_{eff}} \leq 0.3 \tag{4.93}$$

The final result is again a dimensionless number that compares the rate of reaction to the rate of mass transfer to the active sites in the pores. This is a conservative estimate because the concentration gradient depicted by curve 1 is likely to be steeper than the linear plot shown if the rate is very high. Furthermore, for a value of $\eta \geq 0.95$, the inequalities required in equation 4.88 for 1st-order and zero-order reactions are 0.6 and 6, respectively; therefore, if a Weisz-Prater criterion (equation 4.93) of 0.3 or less is used, rates for all reactions with an order of 2 or less should have negligible mass transfer limitations. In addition, if the Weisz-Prater number is greater than 6, pore diffusion limitations definitely exist [20]. This transition region from kinetic to diffusion control occurs frequently around a TOF $\approx 1\,s^{-1}$ for many supported metal catalysts [22].

Illustration 4.4 – Application of the Weisz-Prater Criterion to a Gas-Phase Reaction

The hydrogenation of CO to form CH_4

(1) $$3H_2 + CO \longrightarrow CH_4 + H_2O$$

over the Group VIII metals has been studied [23]. As a specific example, with a feed gas at 1 atm containing a stoichiometric reactant ratio of $H_2/CO = 3$, a 1.75% $Pt/\eta - Al_2O_3$ catalyst gave a rate of 2.3×10^{-7} mole CO s^{-1} $(cm^3\ cat)^{-1}$ at 548 K under differential reactor conditions. Use of the Weisz-Prater parameter

$$N_{W-P} = \Re R_p^2 / C_s D_{eff} \tag{1}$$

requires that R_p, C_s and D_{eff} be known. The catalyst particles passed through a 40-mesh screen, which represents a 0.042 cm opening (see Perry's Chemical Engineering Handbook, for example), thus the largest particles have a maximum radius of $R_p = 0.021$ cm. If one focuses on CO, its bulk concentration, which will be assumed to equal the surface concentration because of the SV used, is:

$$C_s = C_o = N_{CO}/V = P_{CO}/RT$$

$$= (1\,\text{atm})(0.25) \Big/ \left(\frac{82.06\,\text{atm} \cdot \text{cm}^3}{\text{g mole} \cdot \text{K}}\right)(548\,\text{K})$$

$$= 5.6 \times 10^{-6} \text{ mole cm}^{-3} \tag{2}$$

The average velocity for the CO molecules is (equation 4.76):

$$\bar{v} = [8k_BT/\pi m]^{1/2} = \left[\frac{8(1.38 \times 10^{-16}\text{erg} \cdot \text{K}^{-1})(548\,\text{K})}{\pi(28\,\text{amu})(1.66 \times 10^{-24}\text{g} \cdot \text{amu}^{-1})}\right]^{1/2} \tag{3}$$

$$= 6.4 \times 10^4 \text{ cm s}^{-1}$$

Therefore, the mean free path in the gas phase is:

$$\lambda = \frac{1}{\sqrt{2}\pi\sigma^2(N_{CO}/V)} = \frac{RT}{\sqrt{2}\pi\sigma^2 P_{CO}}$$

$$= \frac{\left(\dfrac{82.06\,\text{cm}^3 \cdot \text{atm}}{\text{g mole} \cdot \text{K}}\right)(548\,\text{K})(1\,\text{mole}/6.02 \times 10^{23}\,\text{molecules})}{\sqrt{2}\pi(3.2 \times 10^{-8}\,\text{cm})^2(1\,\text{atm})}$$

$$= 1.6 \times 10^{-5} \text{ cm} \tag{4}$$

where σ is the molecular diameter, and the gas-phase diffusivity is:

$$D = \frac{\bar{v}\lambda}{3} = 1/3(6.4 \times 10^4 \text{ cm s}^{-1})(1.6 \times 10^{-5}\text{ cm}) = 0.35 \text{ cm}^2 \text{ s}^{-1} \tag{5}$$

However, the alumina used had a typical pore diameter of around $d_p = 5$ nm, which is much less than λ(160 nm); consequently, pore diffusion will be dominated by Knudsen diffusion and

$$D_{eff} \cong D_{Kn}$$

$$= \frac{\bar{v}d_p}{3} = 1/3(6.4 \times 10^4 \text{ cm s}^{-1})(50 \times 10^{-8}\text{ cm}) = 0.011 \text{ cm}^2 \text{ s}^{-1} \tag{6}$$

Therefore, the value of the W-P number for CO, with consistent units, is:

$$N_{W-P} = \frac{\Re R_p}{C_S D_{eff}} = \frac{(2.3 \times 10^{-7} \text{ mole CO s}^{-1} \text{ cm}^{-3})(0.021 \text{ cm})^2}{(5.6 \times 10^{-6} \text{ mole CO cm}^{-3})(0.011 \text{ cm}^2 \text{ s}^{-1})}$$

$$= 1.7 \times 10^{-3} \tag{7}$$

which is much less than 0.3. Alternatively, if H_2 is considered, although its transport rate would be expected to be higher, one has

$$\bar{v} = \left[\frac{8\left(1.38 \times 10^{-16}\, \text{erg} \cdot \text{K}^{-1}\right)(548\,\text{K})}{\pi(2\,\text{amu})(1.66 \times 10^{-24}\,\text{g amu}^{-1})} \right]^{1/2} = 2.4 \times 10^5\,\text{cm s}^{-1}, \qquad (8)$$

$$\lambda = \frac{\left(\dfrac{82.06\,\text{cm}^3 \cdot \text{atm}}{\text{g mole} \cdot \text{K}}\right)(548\,\text{K})\left(1\,\text{mole}/6.02 \times 10^{23}\,\text{molecules}\right)}{\sqrt{2}\pi(2.4 \times 10^{-8}\,\text{cm})^2(1\,\text{atm})}$$

$$= 2.9 \times 10^{-5}\,\text{cm}, \qquad (9)$$

and the gas-phase diffusivity is

$$D = 1/3(2.4 \times 10^{-5}\,\text{cm s}^{-1})(2.9 \times 10^{-5}\,\text{cm}) = 2.3\,\text{cm}^2\,\text{s}^{-1} \qquad (10)$$

However, the effective diffusivity is again going to be controlled by the pore dimensions, so

$$D_{\text{eff}} = 1/3(2.5 \times 10^5\,\text{cm s}^{-1})(50 \times 10^{-8}\,\text{cm}) = 4.2 \times 10^{-2}\,\text{cm}^2\,\text{s}^{-1} \qquad (11)$$

and for H_2

$$N_{\text{W–P}} = \frac{(3)(2.3 \times 10^{-7}\,\text{mole } H_2\,\text{s}^{-1}\,\text{cm}^{-3})(0.021\,\text{cm})^2}{(3)(5.6 \times 10^{-6}\,\text{mole } H_2\,\text{cm}^{-3})(0.042\,\text{cm}^2\,\text{s}^{-1})} = 4.3 \times 10^{-4} \qquad (12)$$

which again is far lower than 0.3.

4.2.3.2   Liquid-Phase Reactions

As mentioned previously, the dimensionless Weisz-Prater (W-P) number is convenient to use because $\Re$ is measured directly, C_s can be assumed to be equal to the bulk concentration if all external mass transfer limitations are removed by proper mixing, R_p can be determined experimentally, and D_{eff} can be calculated for vapor-phase reactions quite straightforwardly once the average pore radius is known. However, for liquid-phase reactions using a porous catalyst, including those with a gas-phase reactant (a three-phase system), acquiring a value for the Weisz parameter is not quite so easy for two reasons. First, reasonably accurate values of D_{eff} must be determined for the reactants in the liquid-filled pores and, second, the concentration of the gas in the liquid reactant (if neat) or the reactant/solvent mixture (C_o) must be calculated. This is especially important to do for liquid-phase reactions because they are much more susceptible to mass transport limitations; however, in general, heat transport limitations are of less concern because of the relatively high heat capacities and thermal conductivities of a liquid phase compared to a vapor phase. Let us examine the procedures that exist to allow calculation of C_o and D_{eff} values in liquid systems.

In order to use the W-P criterion successfully, it is important to accurately determine the effective diffusivity of the reactant in a given catalyst system. In the case of liquid-phase reactions, the pores of the catalyst are filled primarily with solvent, if one is used, and the molecular diffusivity of the reactant solute can be several orders of magnitude lower in a liquid-phase system compared to a vapor-phase system. Apart from the decrease in bulk diffusivities, there can also be liquid-phase non-idealities, adsorption phenomena, and other unknown factors influencing the effective diffusivity [24]. If the size of the diffusing molecules is comparable to the pore size, diffusion in the pores is hindered [25–28]. A different situation arises when the diffusing species has a strong affinity for the catalytic surface, which can lead to surface diffusion and migration [2,29]; however, this consideration will not be of importance in liquid-phase reactions [30].

The effective diffusivity, D_{eff}, will depend not only on the relative sizes of the diffusing molecule and the pores, but also on the type of solvent residing in the pores. In order to assess the hindered diffusion of species through a pore structure, it is necessary to possess appropriate pore-size data. The average pore size, which is characterized by the pore radius, r_p, is not an issue with catalysts having relatively uniform pores. If a pore-size distribution exists, a mean pore radius, $\bar{r}_p$, can be estimated using the void volume (per g) $\bar{V}_m$, the specific surface area, A_m, the bulk density of the catalyst pellet, ρ_p, and the catalyst porosity (i.e. the void fraction), ε, as shown below [2]:

$$\bar{r}_p = \frac{2\bar{V}_m}{A_m} = \frac{2\varepsilon}{A_m \rho_p} \tag{4.94}$$

However, a pelletized or extruded catalyst prepared by compacting fine powder typically exhibits a bimodal (macro-micro) pore-size distribution, in which case the mean pore radius is an inappropriate representation of the micropores. There are several analytical approaches and models in the literature which represent pelletized catalysts, but they involve complicated diffusion equations and may require the knowledge of diffusion coefficients and void fractions for both micro- and macro-pores [31]. An easier and more pragmatic approach is to consider the dimensional properties of the fine particles constituting the pellet and use the average pore size of only the micropore system because diffusional resistances will be significantly higher in the micropores than in the macropores. This conservative approach will also tend to underestimate D_{eff} values and provide an upper limit for the W-P criterion.

The topic of the effective diffusivity of a solute in liquid-filled pores has been studied, and modifications to the conventional model for diffusivity by incorporating empirical constants have been provided, such as that shown below [25,28]:

$$D_{eff} = D_b \frac{\varepsilon}{\tau} \{A \exp(-B\lambda)\} \tag{4.95}$$

Here, D_b is the bulk diffusivity, ε and τ are the catalyst porosity and tortuosity, respectively, λ is the ratio of the radius of the diffusing molecule to the pore radius (i.e., $r_{\text{molecule}}/r_{\text{pore}}$), and A and B are empirical constants based on the catalyst and the type of diffusing molecule. Care must be taken not to overlook the phenomenon of hindered diffusion for relatively larger molecules diffusing through micropores of similar dimensions. In a more recent treatment of diffusion of solute molecules in liquid-filled pores, an expression was developed for effective diffusivity which involves one empirical constant (32):

$$D_{\text{eff}} = D_b \frac{(1-\lambda)^2}{1+P\lambda} \tag{4.96}$$

Here, the empirical constant, P, is a fitting parameter determined individually for catalysts using data in the literature that provided sufficiently close fits. For a given liquid-phase system, after choosing an appropriate expression to evaluate the effective diffusivity of a reacting species, the determination of D_b becomes an important consideration. Unless the bulk diffusivity of the particular solute (i.e., reactant)-solvent system at the reaction conditions employed is available from the literature, it has to be estimated from standard formulations based on physical properties of the solute and the solvent. Depending on whether the solute is a gas or a liquid, a suitable expression for diffusivity must be chosen. Some of the common situations and the corresponding expressions to evaluate bulk diffusivity are listed below [33], and the definitions of the quantities involved in equations 4.97–4.103 are listed collectively at the end of this section in Table 4.1.

(a) Diffusivity of a dilute gas solute in a liquid solvent (34):

$$D_{12}^0 = 1.1728 \times 10^{-16} \frac{T\sqrt{\chi M_2}}{\eta_2 V_1^{0.6}} \tag{4.97}$$

(b) Diffusivity of a dilute solute (<10 mol %) in water (35):

$$D_{12}^0 = \frac{8.621 \times 10^{-14}}{\eta_2^{1.14} V_1^{0.589}} \tag{4.98}$$

(c) Diffusivity of a dilute solute (<10 mol %) in any solvent except water (36):

$$D_{12}^0 = 4.4 \times 10^{-15} \frac{T}{\eta_2} \left(\frac{V_2}{V_1}\right)^{1/6} \left(\frac{L_2^{\text{vap}}}{L_1^{\text{vap}}}\right)^{1/2} \tag{4.99}$$

(d) Diffusivity in concentrated binary liquid systems using activity coefficient parameters (37):

TABLE 4.1. Definition of the Terms Involved in Equations
4.97–4.103 (Along with Typical Units)

D_{ij}^0 = diffusivity at infinite dilution of i in j [cm^2 s^{-1}]
D_{ij} = diffusivity of i in the concentrated binary mixture [cm^2 s^{-1}]
T = system temperature [K]
χ = solvent association parameter
M_i = molecular weight of component i
η_i = viscosity of component i [Pa s]
x_i = mole fraction of component i
V_i = molar volume of component i at normal boiling point [m^3 kmol^{-1}]
L_i^{vap} = enthalpy of vaporization of component i at normal boiling point
    [J kmol^{-1}]
α_{12} = thermodynamic correction term
γ_i = activity coefficient of component i
Subscripts: 1 = solute
            2, 3 = solvents
            m = mixture

$$D_{12} = \left(x_1 D_{21}^0 + x_2 D_{12}^0\right)\alpha_{12} \qquad (4.100)$$

$$\text{where, } \alpha_{12} = 1 + \frac{d(\ln \gamma_1)}{d(\ln x_1)} \qquad (4.101)$$

(e) Diffusivity in concentrated binary liquid systems using viscosity
    data (38):

$$D_{12} = \frac{\left(D_{21}^0 \eta_1\right)^{x_1}\left(D_{12}^0 \eta_2\right)^{x_2}}{\eta_m} \qquad (4.102)$$

The reaction under consideration may involve a solvent and/or one or
more liquid-phase products, thus making it a multi-component diffusion
system. In such cases, D_b represents the solute diffusivity in the
liquid mixture including the products. To simplify the effort to make a
reasonable estimate of D_b, the relatively insignificant components may
be neglected. For example, the following expression can then be used to
calculate the diffusivity of a solute, gas or liquid, in 2-solvent liquid system
(39):

$$\ln\left(D_{1m}\eta_m^{0.5}\right) = x_2 \ln\left(D_{12}^0 \eta_2^{0.5}\right) + x_3 \ln\left(D_{13}^0 \eta_3^{0.5}\right) \qquad (4.103)$$

Finally, in gas-liquid systems, Henry's law can be used to calculate or to
estimate the concentration of the gas in the solvent or the reacting liquid
mixture.

An example of the utilization of these equations to calculate D_{eff} and C_o
values in a real system so that W-P numbers could be estimated is provided
in Illustration 4.7. These same techniques can also be used to calculate D_{eff}
values for use in a Thiele modulus, of course.

Illustration 4.5 – Calculating D_{eff} and C_o Values and Evaluating the W-P Criterion for a Liquid-Phase Reaction

Due to the presence of a carbonyl (C=O) bond conjugated with a C=C double bond, plus an isolated C=C double bond, the hydrogenation of citral (3,7-dimethyl-2,6-octadienal) has a rather complex reaction network involving 6–8 important intermediates and a number of series-parallel reactions [40–45], each of which represents the addition of one mole of dihydrogen, as shown in Fig. 4.10. Although the W-P criterion was originally derived for a much less complicated catalytic reaction, it can still be applied to this type of chemical system by examining the relative diffusivity of each reactant (citral and hydrogen) in the liquid-filled pores. The reactant with the lower diffusivity cannot be automatically assumed to be the diffusion-limiting species (if one does exist) because the liquid-phase concentration and the relative rate of consumption are also considered when evaluating the W-P number. The reaction data examined here were obtained from citral hydrogenation runs in various solvents performed at 21.1 atm pressure and 373 K conducted by Mukherjee using a silica-supported Pt catalyst [46]. The experimental details of the semi-batch reactor system, composition analysis,

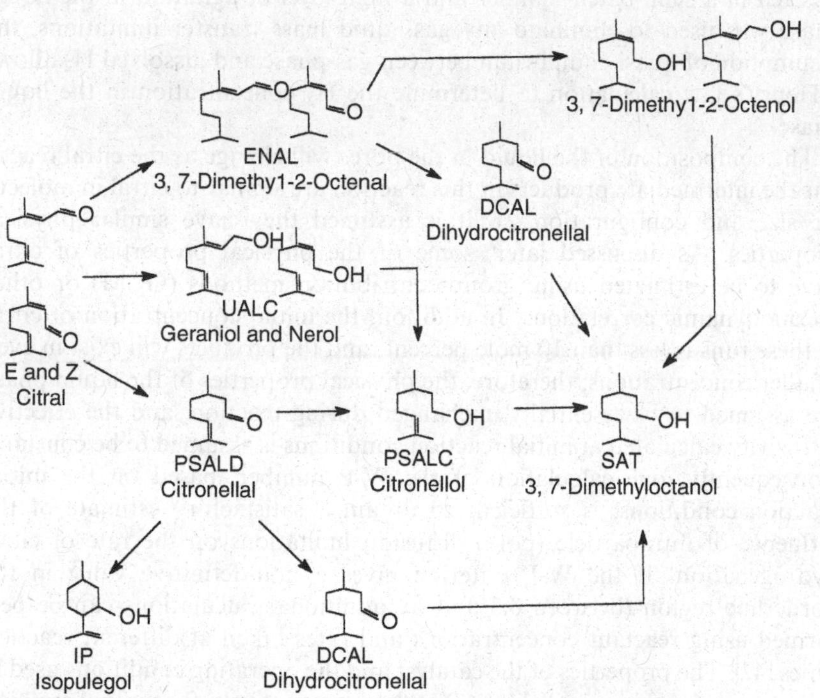

FIGURE 4.10. Reaction network for citral hydrogenation. (Reprinted from ref. 8, copyright © 2001, with permission from Elsevier)

and other reaction parameters can be found in the studies conducted by Singh and Vannice [6–8,43–45]. Their experiments verified the absence of external and internal mass transfer limitations using the Madon-Boudart technique, which is discussed next in section 4.2.4.1. Singh and Vannice studied citral hydrogenation with n-hexane as the solvent while in the latter study, seven other solvents were used [46].

Based on the reaction chemistry, questions may arise regarding variations in the parameters in the W-P criterion as the reaction progresses in a batch or semi-batch reactor. The rate of reaction based on citral disappearance decreases because the reaction is carried out in a batch mode relative to citral, whose concentration drops as the reaction proceeds; consequently, the composition of the liquid inside the pores changes, which could alter the effective diffusivity. Such issues within a reaction system must be considered and an appropriate choice of modeling the intraparticle diffusion should be made to obtain a reasonable simplification to acquire a single value for the W-P number. The initial reaction rate, which is the highest during the course of the reaction, is chosen for $\Re$. For the surface concentration term, C_s, the corresponding bulk value at initial conditions is chosen. The hydrogen concentration in the liquid-phase medium is determined by the partial pressure of H_2 and the gas-liquid mass transfer rate; however, because H_2 was fed in a semi-batch manner and a high level of agitation in the liquid phase was used to eliminate any gas-liquid mass transfer limitations, the assumption of quasi-equilibrium between gas-phase and dissolved H_2 allows a Henry's law calculation to determine the H_2 concentration in the liquid phase.

The composition of the liquid in the pores will change as the citral reacts, but the intermediate products in this reaction are similar to citral in molecular size and configuration, so it is assumed they have similar physical properties. As discussed later, some of the physical properties of citral have to be estimated using group-contribution methods (GCM) or other thermodynamic correlations. In addition, the initial concentration of citral in these runs is less than 10 mole percent, and the products will exist in even smaller concentrations; therefore, the physical properties of the liquid phase are assumed to be essentially unchanged during reaction, and the effective diffusivity calculated at initial reaction conditions is assumed to be constant. Consequently, one calculation of the W-P number, based on the initial reaction conditions, is sufficient to obtain a satisfactory estimate of the influence of intraparticle (pore) diffusion limitations on the rate of citral hydrogenation. If the W-P criterion gives a non-definitive value in the borderline region (between 0.3 and 6), additional calculations can be performed using reactant concentrations and rates taken at different reaction times [47]. The properties of the catalyst and the operating conditions used in this reaction are given in Tables 4.2 and 4.3, respectively. The initial reaction rate observed with different solvents was below a TOF of $1\,s^{-1}$, or a rate of $2.5 \times 10^{-6}\ mol\,s^{-1}\,(cm^3\,catalyst)^{-1}$, at the conditions mentioned in Table 4.3.

TABLE 4.2. Catalyst Properties (ref. 46)

Catalyst	3% Pt/SiO$_2$
Preparation	Ion-exchange method
Pretreatment	Reduced in-situ in flowing H$_2$ at 673K for 75 min
Metallic dispersion	Pt$_s$/Pt$_{total}$ = 1.0
Particle radius	R$_p$ = 50 µm
Pore radius*	r$_p$ = 7 nm

*Based on the pore size of Grade 57 Grace-Davison silica

TABLE 4.3. Reaction Conditions for Citral Hydrogenation (ref. 46)

Temperature	373 K
Pressure	21.1 atm
Citral concentration	0.5 mol/L
Agitation	1000 rpm
Total liquid volume	60 cm^3
Solvents used	n-Amyl Acetate, Ethanol, Ethyl Acetate, Cyclohexanol Cyclohexane, n-Hexane, p-Dioxane, THF (tetrahydrofuran)
Max. initial rate	2.5 × 10^{-6} mol s^{-1}(cm cat)$^{-3}$

The principal quantity that needs to be determined is the effective diffusivity. Because there are two diffusing reactants in this system, namely citral and H$_2$, it is necessary to compute the diffusivities of each in the liquid-filled pores. As mentioned earlier, Ternan has derived an expression for the effective diffusivity based only on λ (i.e., r$_{molecule}$/r$_{pore}$) and a single fitting parameter (32). This author applied two correction factors to the bulk diffusivity – one for a concentration effect and one for a pore wall effect – to arrive at the effective diffusivity. The pore cross-sectional region near the pore wall corresponding to (r$_{pore}$ − r$_{molecule}$) was assumed to be unavailable to solute molecules, although smaller solvent molecules could be assumed to utilize the excluded region, thus causing a decreased concentration of the solute in the pore compared to its concentration in the bulk liquid immediately outside the pore. This geometric correction is addressed via the "concentration effect". The latter correction for a "pore wall effect" considers the viscosity variation near the pore wall caused by the force field of the wall. The significance of the viscosity can be realized from the fact that in any of the basic equations for bulk molecular diffusivity for a solute-solvent pair at a given temperature, D_b depends only on the viscosity, η, of the medium. This is reflected by the Stokes-Einstein equation describing the relationship between molecular diffusivity and solvent viscosity:

$$\frac{D\eta}{T} = \text{constant} \qquad (1)$$

The final form of the effective diffusivity expression derived by Ternan is given by equation 4.96:

$$D_{\text{eff}} = D_b \frac{(1 - \lambda)^2}{1 + P\lambda} \qquad (2)$$

The fitting parameter, P, was calculated by Ternan to be 16.26 based on data representing measured liquid diffusivities of organics in a silica-alumina catalyst, as reported by Satterfield et al. [28]. In the absence of diffusivity data for the silica used in this study, the above correlation with $P = 16.3$ was utilized to compute effective diffusivities. If it is assumed that the properties of the liquid inside the pores and the tortuosity factor remain unchanged, equation 2 shows that D_{eff} is proportional to D_b, which is the molecular diffusivity of the solute in the reaction medium. Thus, the effect of the solvent will manifest itself through the D_b values.

As an example, let us now calculate the D_{eff} and C_o values with ethanol (EtOH) as the solvent. First, H_2 will be considered as the diffusing reactant, and the diffusivity of H_2 in citral and the solvent can be estimated using equation 4.97, which is based on the method of Wilke and Chang [34], because the mole fraction of H_2 in either liquid is very low, i.e.,

$$D_{\text{H}_2/\text{citral}} (\text{m}^2\text{s}^{-1}) = 1.1728 \times 10^{-16} \frac{T \sqrt{\chi_{\text{citral}} M_{\text{citral}}}}{\eta_{\text{citral}} V_{\text{H}_2}^{0.6}} \qquad (3)$$

and

$$D_{\text{H}_2/\text{EtOH}} (\text{m}^2\ \text{s}^{-1}) = 1.1728 \times 10^{-16} \frac{T \sqrt{\chi_{\text{EtOH}} M_{\text{EtOH}}}}{\eta_{\text{EtOH}} V_{\text{H}_2}^{0.6}} \qquad (4)$$

The constants in the above equations are represented in units to give m^2s^{-1} for diffusivity, and care must be taken to use consistent units for the quantities involved. The properties of the solvents used in the calculations are given in Table 4.4. It should be noted that the H_2 concentration in the

TABLE 4.4. Physical Properties of Citral and the Solvents [54]

Liquid	Mol. wt.	V_{mol}@T_b ($\text{m}^3\ \text{kmol}^{-1}$)	χ^a	L^{vap}@$T_b{}^b$ ($\text{J kmol}^{-1} \times 10^{-7}$)	η@373 K ($\text{Pa s} \times 10^4$)	$C_{\text{S,H}_2}{}^c$ ($\text{mol cm}^{-3} \times 10^5$)
Citral	152	0.171^d	1	4.41^d	4.22^d	—
n-Amyl Acetate	130	0.175	1	3.85	3.57	9.53
Ethanol	46	0.063	1.5	3.94	3.32	8.38
Ethyl Acetate	88	0.106	1	3.22	2.12	11.5
Cyclohexanol	100	0.123	1	4.59	20.2	6.97
Cyclohexane	84	0.117	1	2.99	3.05	9.06
n-Hexane	86	0.140	1	2.91	1.58	12.5
p-Dioxane	88	0.094	1	3.43	3.90	6.89
THF	72	0.086	1	3.03	2.34	10.2

a) Ref (33)
b) Latent heat of vaporization at normal boiling point (T_b)
c) At 373 K, 21.1 atm total pressure, and 0.5M citral
d) Computed as described in this example

liquid phase is only on the order of $0.1 \, \text{mmol cm}^{-3}$, thus H_2 may be neglected as a constituent in the bulk phase for all volumetric purposes as the molar volume of H_2 at its normal boiling point is $0.0286 \, \text{m}^3 \, \text{kmol}^{-1}$ [48]. At a reaction temperature of 373 K, the diffusivity of H_2 in citral or ethanol is calculated to be 1.1×10^{-4} or $9.3 \times 10^{-5} \, \text{cm}^2 \, \text{s}^{-1}$ from equations 3 and 4, respectively. The effective binary diffusion coefficient for H_2 in the mixture can now be computed using equation 4.103 after its rearrangement [39], i.e.,

$$D_{H_2/\text{mixt}} = \frac{\left(D_{H_2/\text{citral}} \, \eta_{\text{citral}}^{0.5}\right)^{x_{\text{citral}}} \left(D_{H_2/\text{EtOH}} \, \eta_{\text{EtOH}}^{0.5}\right)^{x_{\text{EtOH}}}}{\eta_{\text{mixt}}^{0.5}} \tag{5}$$

The viscosity of the mixture required in equation 5 may be computed from the relationship [33]:

$$\eta_{\text{mixt}} = \eta_{\text{citral}}^{x_{\text{citral}}} \, \eta_{\text{EtOH}}^{x_{\text{EtOH}}} \tag{6}$$

A value of $9.3 \times 10^{-5} \, \text{cm}^2 \, \text{s}^{-1}$ was obtained for the diffusivity of H_2 in the mixture, which, as expected, is essentially that of H_2 in ethanol because of the small mole fraction of citral ($x_{\text{citral}} = 0.032$).

Obtaining physical properties for an uncommon chemical such as citral has always been a challenge and information about it is scarce, but fortunately, there are methods available to calculate thermodynamic properties [33,49–52]. The data prediction manual compiled by Danner and Daubert is extremely useful for the estimation of diffusivity and thermal or physical properties of compounds in the absence of experimental data [33]. In this example, a number of properties pertaining to citral, such as critical temperature (T_c), critical pressure (P_c), viscosity (η), and enthalpy of vaporization (L^{vap}) had to be estimated, and these calculated values are listed in Table 4.5 while details of the procedures for their estimation are given elsewhere [46]. The values for the surface concentration of H_2 in the last column of Table 4.4 are those given by the H_2 solubility in the medium at the mentioned conditions, which was determined based on Henry's law constants for H_2 in citral and the respective solvents. These Henry's law constants were either compiled from the literature or computed using thermodynamic techniques based on the Soave-Redlich-Kwong (SRK) equation of state [53].

TABLE 4.5. Calculated Physical Properties of Citral (Ref. 46).

Quantity	Calculated value	References
T_c	699 K	49,55
P_c	22.6 atm	49,55
η	$\log_{10} \eta = 509.12(T^{-1} - 3.42 \times 10^{-3}) - 3.0 = 4.2 \times 10^4 \, \text{P}_a \, \text{s@373 K}$	33,51
ρ_{sat}	$\rho_{\text{sat}} = (0.0599)0.2334^{[-1(1-T/Tc)^{2/7}]} = 0.83 \, \text{g cm}^{-3}\text{@373 K}$	33,50
L^{vap}	$L^{\text{vap}} = 4.55 \times 10^7 \, \text{J kmol}^{-1}\text{@}T_b = 501 \, \text{K}$	33,52
δ	$\delta_{298\,\text{K}} = 18.3 \, \text{J}^{0.5} \, \text{cm}^{-1.5}$	56

In the second part of this analysis, citral is considered to be the diffusing component. To estimate the binary diffusivity of citral in ethanol, equation 4.99, which is applicable to a dilute solute ($x_{citral} < 0.1$) in a non-water solvent, is used to estimate diffusivity [36]:

$$D_{citral/mixt} \cong D_{citral/EtOH} \left(m^2 \ s^{-1}\right)$$

$$= 4.4 \times 10^{-15} \frac{T}{\eta_{EtOH}} \left(\frac{V_{EtOH}}{V_{citral}}\right)^{1/6} \left(\frac{L_{EtOH}^{vap}}{L_{citral}^{vap}}\right)^{1/2}$$

With the values in Table 4.4, the value of $D_{citral/mixt}$ turns out to be $3.9 \times 10^{-5} \ cm^2 \ s^{-1}$. With these values for the bulk diffusivities of H_2 and citral in the reaction mixture, the effective diffusivity of each of the diffusing species in the catalyst pores can be computed using equation 2, and the results are tabulated in Table 4.6 along with additional relevant information.

The effective diffusivities for citral and H_2 are significantly lower than the comparable bulk diffusivities; consequently, the correction for hindered diffusion in the pores is significant. The greater correctional effect for citral diffusivity could be expected on account of its size. The D_{eff} values also show that H_2 diffusivity in the pores is almost four times higher than that of citral; nevertheless, it should not be automatically inferred that the slower diffusion of the citral molecules results in citral being the more likely limiting reactant. This is because the W-P criterion contains not only the diffusivity of the reactant species, but also its rate of consumption and its concentration at the surface of the catalyst particle, as shown in equation 4.93. The surface concentration of H_2 in a solution of 1M citral in ethanol is that given by the hydrogen solubility in the solution at 373 K, i.e., $C_{S,H_2} = 83.8 \ \mu mol \ cm^{-3}$; consequently, application of the W-P criterion for each reactant gives:

$$N_{W-P}|_{H_2} = \frac{\left(2.5 \times 10^{-6} \frac{mol}{cm^3 s}\right)(0.005 \ cm)^2}{\left(8.4 \times 10^{-5} \frac{mol}{cm^3}\right)\left(7.0 \times 10^{-5} \frac{cm^2}{s}\right)} = 0.011 \ll 0.3 \qquad (8)$$

and

$$N_{W-P}|_{citral} = \frac{\left(2.5 \times 10^{-6} \frac{mol}{cm^3 s}\right)(0.005 \ cm)^2}{\left(5 \times 10^{-4} \frac{mol}{cm^3}\right)\left(1.8 \times 10^{-5} \frac{cm^2}{s}\right)} = 0.0070 \ll 0.3 \qquad (9)$$

The value of N_{W-P} for either reactant is at least an order of magnitude less than 0.3, thus each satisfies the condition assuring the absence of significant

TABLE 4.6. Effective Diffusivities of Citral and H_2 in Ethanol at 373 K (Ref. 46)

Component	$r_{molecule}$(nm)	$\lambda = \frac{r_{molecule}}{r_{pore}}$	D_b(cm^2 s^{-1} × 10^5)	D_{eff}(cm^2 s^{-1} × 10^5)
Citral	0.39	0.056	3.8	1.8
Hydrogen	0.12	0.017	9.3	7.0

pore diffusional limitations during citral hydrogenation in an ethanol solvent. Surprisingly, the $N_{W\text{-}P}$ value with respect to H_2 is higher than that for citral, indicating a higher probability that the intraparticle diffusion of H_2, rather than citral, could inhibit the rate of reaction. This demonstrates the importance of the surface concentration term in the evaluation of the W-P parameter because it provides the driving force for the diffusion into the pores, and the transport rate is D_{eff} multiplied by C_s/R_p. In these calculations the same reaction rate was used for both dihydrogen and citral, which is justified because the initial (and maximum) rate of reaction primarily involves the conversion of citral to geraniol, nerol and citronellal; hence, the rate of citral disappearance equals the rate of dihydrogen consumption. If significant secondary reactions were to occur in series to form products like citronellol and 3,7-dimethyloctanol within the time period associated with the first calculation, the molar rate of hydrogen consumption would exceed that of citral, which would increase the hydrogen-based $N_{W\text{-}P}$ value and further accentuate the possibility that H_2 is more likely to be the reactant to induce diffusion-limiting constraints on the rate. Results for other solvents are provided elsewhere [46].

4.2.4   Other Criteria to Verify the Absence of Mass and Heat Transfer Limitations (Madon-Boudart Method)

To decrease and remove external gradients, the velocity of fluid (gas or liquid) relative to the catalyst particles must be increased, thus in a liquid-phase slurry reaction within a batch or semi-batch reactor, the mixing parameter (such as stirring speed) is increased until no further increase in rate is observed. Above this speed no external gradients are assumed to exist. External gradients can occur in these systems and they can have a significant effect upon both activity and selectivity [57–60]. In a fixed-bed catalytic reactor, external mass transport gradients are assumed negligible if the reaction rate remains constant as the volumetric flow rate, V_o, is varied but the space time, V_{cat}/V_o, is kept constant by either decreasing the amount of catalyst or by diluting it with an inert material with similar flow properties. However, under some conditions, such as a low Reynolds number ($Re = d_p v_f \, \rho_f/\eta_f$, where d_p is the particle diameter while the flow velocity, v_f, density, ρ_f, and viscosity η_f, relate to the fluid), this latter method may not be very sensitive [61]. To check for the influence of pore diffusion, historically the catalyst particle size has been decreased to determine the effect on rate, with an increase in rate indicating mass transfer limitations. However, there is a realistic lower limit on particle size experimentally, and if the average pore size is much smaller than the smallest particle size, which is frequently the case, then mass transfer limitations within the pores can still remain.

In the kinetic control regime (where the overall effectiveness factor $\eta = 1$), the rate is directly proportional to the concentration of active sites, L, which is incorporated into the rate constant. In the regime of internal (pore) diffusion control, the rate becomes proportional to $L^{1/2}$, and when external diffusion controls the rate there is no influence of L, i.e., there is a zero-order dependence on L. This can be seen by examining equations 4.47 and 4.68. This observation led to the proposal by Koros and Nowak to test for mass transfer limitations by varying L [62]. This concept was subsequently developed further by Madon and Boudart to provide a test that could verify the absence of any heat and mass transfer effects as well as the absence of other complications such as poisoning, channeling and by-passing [63].

The Madon-Boudart technique is summarized as follows. If a rate constant, $k_o(s^{-1})$ is written in terms of a turnover frequency (TOF), then the specific rate constant for the catalyst, k, will be:

$$k = L \, k_o \tag{4.104}$$

Keep in mind that the site density, L, can be expressed either per unit surface area or per g catalyst, typically the latter. It has been shown in the preceding paragraph that:

$$\Re \alpha \, L^s \tag{4.105}$$

and this proportionality can be expressed quantitatively using a Thiele modulus, ϕ, as:

$$\Re = \eta \Re_o = \phi^p L \, k_o C_o^n \tag{4.106}$$

thus:

$$\Re \alpha \, L^{1+p/2} = L^s \tag{4.107}$$

where the effectiveness factor $\eta = \phi^p$ and $s = 1 + p/2$.

In the absence of any mass transfer limitations, $\eta = 1$, $p = 0$ and $s = 1$; however, in the presence of the most severe instances of pore diffusion control, ϕ is large and $\eta = 1/\phi$ so $p = -1$ and $s = 1/2$. If L is now varied throughout the pore structure of the catalyst and $\ln \Re$ is plotted vs. $\ln L$, then the slope equals s. If the slope equals unity, then $\eta = 1$ and the TOF will be constant for all L. However, if the slope decreases significantly from unity, then either internal ($s = 1/2$) or external mass transfer gradients are controlling the reaction rate. If a slope of unity is attained at two different temperatures, then the system is isothermal and no temperature gradients exist [63]. An example of the latter situation is shown in Figure 4.11 for the liquid-phase hydrogenation of cyclohexene [63], and slopes near unity were also observed during the liquid-phase hydrogenation of citral [6–8].

One caveat must be mentioned here if this technique is applied to catalysts with a dispersed active phase, such as supported metal catalysts, i.e., the

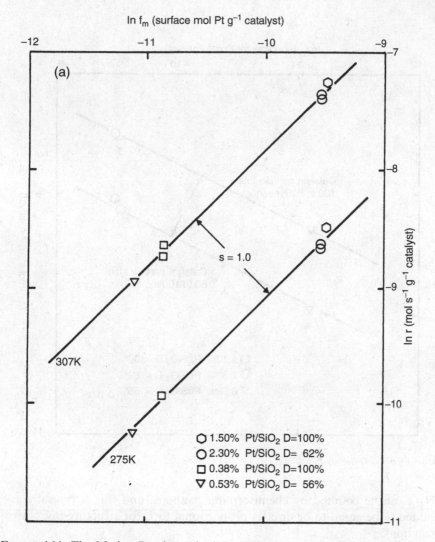

FIGURE 4.11. The Madon-Boudart criterion applied in a study of the liquid-phase hydrogenation of cyclohexene on Pt/SiO$_2$ catalysts. Reaction conditions: $P_{H_2} = 101.3$ kPa, solvent $= 20$ cm^3 cyclohexane. (a) Catalyst particle size > 200 mesh, 2 cm^3 cyclohexene added at 275 K and 0.5 cm^3 cyclohexene added at 307K; (b) 0.5 cm^3 cyclohexene added. (Reprinted from ref. 63, copyright © 1982, with permission from American Chemical Society)

reaction rate may not be proportional to the metal surface area if the reaction is structure sensitive. In this case, the experimenter must be careful to keep the metal dispersion (N_{M_s}/N_{M_t}) constant or nearly constant as the metal loading is varied. Of course, the total number of surface metal atoms,

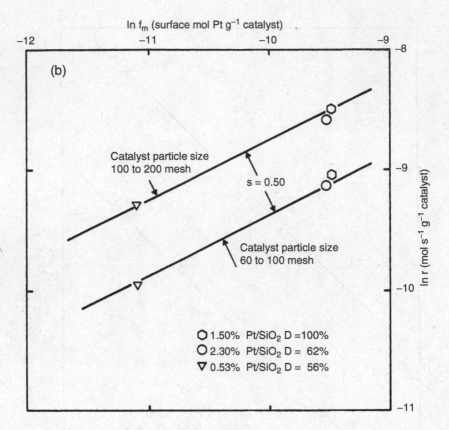

FIGURE 4.11. *continued*

N_{M_s}, can be counted by chemisorption methods, and this is typically assumed to be equal to or directly proportional to L for structure-insensitive reactions.

4.2.5   Summary of Tests for Mass and Heat Transfer Effects

So, the following is a summary of the methods at our disposal to check for mass and heat transfer effects.

1. With slurry-phase systems, the mixing capability is increased until the reaction rate is independent of mixing – external gradients are then assumed to be gone. With gas/solid reactions in a CSTR, the internal recycle rate or the spinning speed of the catalyst basket can be increased to eliminate external gradients [9–11]. A Damköhler number can then be estimated to see if $\bar{\eta}$ is sufficiently close to unity.

2. In a fixed-bed PFR, the flow rate is varied at constant space time and if the rate remains constant, external mass transfer effects are assumed to be unimportant; however, this test can become insensitive at low Reynolds number [61].

3. The catalyst particle size can be varied, and an increase in rate as the particles become smaller shows that internal (pore) diffusion is affecting the rates, assuming that external mass transfer limitations have been removed. Diffusion control may remain in even the smallest particles if the pore diameters are too small.

4. Rates can be measured, the catalysts can be characterized, and the effective diffusivity can be calculated so that a value for the Thiele modulus can be determined (see equation 4.82), and from this a value for η can be obtained (equation 4.66). If η is sufficiently close to unity, the absence of internal diffusion control is indicated.

5. The same information required in procedure 4 can be used directly to evaluate the Weisz-Prater criterion. Values below 0.3 imply $\eta > 0.95$ and the absence of significant pore diffusion limitations for essentially all reactions of interest (reaction order n for $0 \leq n \leq 2$), while values above 6 indicate the strong influence of pore diffusion. It must be stressed that this criterion should be viewed as an order of magnitude estimate, thus values much less than 0.3 are desirable, and values between 0.3 and 6 must be interpreted with caution because an influence due to pore diffusion cannot be ruled out. In addition, if significant product inhibition occurs, the W-P criterion becomes inapplicable in the form presented in this chapter; however, it can be modified to again provide a useful test for pore diffusion effects [64].

6. Finally, one has the Madon-Boudart technique in which the number of active sites is varied by changing the metal loading or, for structure-insensitive reactions, by altering the metal dispersion or perhaps by poisoning sites. In contrast, the metal dispersion should be maintained constant while L is varied for structure-sensitive reactions. A slope of unity from a plot of $\ln \Re$ vs. $\ln L$, i.e., a constant TOF, verifies the absence of ANY mass transfer limitations.

To verify the absence of any heat transfer effects, the Madon-Boudart method can also be utilized. If plots of $\ln \Re$ vs. $\ln L$ are obtained at two or more temperatures, and the slopes are unity at each temperature, then thermal gradients, as well as concentration gradients and other artifacts such as poisoning, can also be ruled out [63]. If the reaction order is known, the criterion in equation 4.73 can also be used to check for isothermal operation.

Other possibilities that can complicate the measurement of accurate rate data can be enumerated, and they have been classified as "parasitic phenomena" by Boudart and Djéda-Mariadassou [22]. These various topics are addressed in this reference and the reader is invited to read about them there.

References

1. C. G. Hill, Jr., "An Introduction to Chemical Engineering Kinetics and Reactor Design", Wiley, NY, 1977.
2. J. M. Smith, "Chemical Engineering Kinetics", 3rd ed., McGraw Hill, NY, 1987.
3. O. Levenspiel, "Chemical Reaction Engineering", 3rd ed., Wiley, NY, 1999.
4. H. S. Fogler, "Elements of Chemical Reaction Engineering", 3rd ed., Prentice Hall, Upper Saddle River, NJ, 1999.
5. J. B. Butt, "Reaction Kinetics and Reactor Design", 2nd ed., Marcell Dekker, NY, 2000.
6. U. K. Singh and M. A. Vannice, *AIChE J.* 45 (1999) 1059.
7. U. K. Singh and M. A. Vannice, *J. Catal.* 191 (2000) 165.
8. U. K. Singh and M. A. Vannice, *Appl. Catal. A.* 213 (2001) 1.
9. D. G. Tajbl, J. B. Simons and J. J. Carberry, *Ind. & Eng. Chem. Fund.* 5 (1966) 171.
10. C. O. Bennett, M. B. Cutlip and C. C. Yang, *Chem. Eng. Sci.* 27 (1972) 2255.
11. J. M. Berty, *Chem. Eng. Prog.* 70 (1974) 78.
12. J. J. Carberry, "Chemical and Catalytic Reaction Engineering", McGraw-Hill, NY, 1976.
13. J. M. Thomas and W. J. Thomas, "Principles and Practice of Heterogeneous Catalysis", VCH, Weinheim, 1997.
14. E. W. Thiele, *Ind. Eng, Chem.* 31 (1939) 916.
15. A. Wheeler, *Adv. Catal.*, Vol. III, p. 249, Academic Press, NY, 1951.
16. A. Wheeler, in "Catalysis", P. H. Emmett, Ed., Vol. 2, Reinhold, New York, 1955.
17. D. E. Mears, *Ind. & Eng. Chem., Proc. Des. Develop* 10 (1971) 541.
18. D. E. Mears, *J. Catal.* 20 (1971) 127.
19. D. E. Mears, *J. Catal.* 30 (1973) 283.
20. P. B. Weisz, *Z. Physik. Chem. NF* 11 (1957) 1.
21. P. B. Weisz and C. D. Prater, *Adv. Catal.* 6 (1954) 143.
22. M. Boudart and G. Djéga-Mariadassou, "Kinetics of Heterogeneous Catalytic Reactions", Princeton Press, Princeton, NJ, 1984.
23. M. A. Vannice, *J. Catal.* 37 (1975) 449, 462.
24. P. A. Ramachandran and R. V. Chaudhari, *Three-Phase Catalytic Reactors*, Gordon and Breach Science, New York, 1983.
25. A. Chantong and F. E. Massoth, *AIChE J.* 29 (1983) 725.
26. C. N. Satterfield and J. R. Katzer, *Adv. Chem. Ser.* 102 (1971) 193.
27. B. D. Prasher and G. A. Gabrill, *AIChE J.* 24 (1978) 1118.
28. C. N. Satterfield, C. K. Colton and W. H. Pitcher Jr., *AIChE J.* 19 (1973) 628.
29. C. N. Satterfield, *Mass Transfer in Heterogeneous Catalysis*, MIT Press, Cambridge, MA, 1970.
30. D. N. Miller and R. S. Kirk, *AIChE J.* 8 (1962) 183.
31. N. Wakao and J. M. Smith, *Chem. Eng. Sci.* 17 (1962) 825.
32. M. Ternan, *Can. J. Chem. Eng.* 65 (1987) 244.
33. R. P. Danner and T. E. Daubert, *Manual for Predicting Chemical Process Design Data*, Design Institute for Physical Property Data, AIChE, New York, 1983.
34. C. R. Wilke and P. Chang, *AIChE J.* 1 (1955) 264.
35. W. Hayduk and H. Laudie, *AIChE J.* 20 (1974) 611.
36. C. J. King, L. Hsueh and K. Mao, *J. Chem. Eng. Data* 10 (1965) 348.

37. C. S. Caldwell and A. L. Babb, *J. Phys. Chem.* 60 (1956) 51.
38. J. Leffler and H. T. Cullinan, *Ind. Eng. Chem. Fund.* 9 (1970) 84.
39. Y. P. Tang and D. M. Himmelblau, *AIChE J.* 11 (1965) 54.
40. M. A. Aramend'ÿa, V. Borau, C. Jimenez, J. M. Marinas, A. Porras, and F. Urbano, *J. Catal.* 172 (1997) 46.
41. B. Bachiller-Baeza, A. Guerrero-Ruiz, P. Wang and I. Rodrÿguez-Ramos, *J. Catal.* 204 (2001) 450.
42. T. Salmi, P. Maki-Arvela, E. Toukoniitty, A. K. Neyestanaki, L. Tiainen, L. Lindfors, R. Sjoholm and E. Laine, *Appl. Catal. A: Gen.* 196 (2000) 93.
43. U. K. Singh and M. A. Vannice, *J. Catal.* 199 (2001) 73.
44. U. K. Singh, M. N. Sysak and M. A. Vannice, *J. Catal.* 191 (2000) 181.
45. U. K. Singh and M. A. Vannice, *J. Molec. Catal. A* 163 (2000) 233.
46. S. Mukherjee, "Solvent Effects in Catalytic Reactions", Ph.D. Thesis, The Pennsylvania State University, 2005.
47. G. C. M. Colen, G. Van Dujin and H. J. Van Oosten, *Appl. Catal.* 43 (1988) 339.
48. T. E. Daubert and R. P. Danner, *Physical and Thermodynamic Properties of Pure Chemicals: Data Compilation*, Hemisphere Pub., New York, 1989.
49. A. L. Lydersen, *Estimation of Critical Properties of Organic Compounds by the Method of Group Contributions*, Univ. Wis. Eng. Exp. Stn. Rep., vol. 3, 1955.
50. C. F. Spencer and R. P. Danner, *J. Chem. Eng. Data* 17 (1972) 236.
51. D. van Velzen and R. L. Cardozo, *I & E C Fund.* 11 (1972) 20.
52. B. I. Lee and M. G. Kesler, *AIChE J.* 21 (1975) 510.
53. J. M. Moysan, M. J. Huron, H. Paradowski and J. Vidal, *Chem. Eng. Sci.* 38 (1983) 1085.
54. C. L. Yaws, *Chemical Properties Handbook: Physical, Thermodynamic Environmental, Transport, Safety and Health Related Properties for Organic and Inorganic Chemicals,* McGraw-Hill Handbooks, McGraw-Hill, New York, 1999.
55. C. F. Spencer and T. E. Daubert, *AIChE J.* 19 (1973) 482.
56. J. H. Hildebrand and R. L. Scott, "The Solubility of Nonelectrolytes", ACS Monograph Series #17, 3rd ed., Reinhold, N.Y., 1950.
57. U. K. Singh, R. N. Landau, Y. Sun, C. LeBlond, D. G. Blackmond, S. Tanielyan and R. L. Augustine, *J. Catal.* 154 (1995) 91.
58. Y. Sun, R. N. Landau, J. Wang, C. LeBlond and D. G. Blackmond, *J. Am. Chem. Soc.* 118 (1996) 1348.
59. Y. Sun, J. Wang, C. LeBlond, R. N. Landau and D. G. Blackmond, *J. Catal.* 16 (1996) 759.
60. Y. Sun, J. Wang, C. LeBlond, R. A. Reamer, J. Laquidara, J. R. Sowa, Jr. and D. G. Blackmond, *J. Organomet. Chem.* 548 (1997) 65.
61. R. P. Chambers and M. Boudart, *J. Catal.* 6 (1966) 141.
62. R. M. Koros and E. J. Nowak, *Chem. Eng. Sci.* 22 (1967) 470.
63. R. J. Madon and M. Boudart, *Ind. & Eng. Chem. Fund.* 21 (1982) 438.
64. E. E. Petersen, *Chem. Eng. Sci.* 20 (1965) 587.
65. N. Krishnankutty and M. A. Vannice, *J. Catal.* 155 (1995) 312.
66. N. Krishnankutty and M. A. Vannice, *J. Catal.* 155 (1995) 327.
67. M. C. J. Bradford and M. A. Vannice, *Cat. Rev. – Sci. & Eng.* 41 (1999) 1.
68. M. C. J. Bradford, "CO_2 Reforming of CH_4 over Supported Metals", Ph.D. Thesis, The Pennsylvania State University, 1997.

Problem 4.1

The following observed rate data have been reported for Pt-catalyzed oxidation of SO_2 at 763K obtained in a differential fixed-bed reactor at atmospheric pressure and containing a catalyst with a bed density of 0.8 g cm^{-3} [2]. The catalyst pellets were 3.2 by 3.2 mm cylinders, and the Pt was superficially deposited upon the external surface. If the void volume of the catalyst bed is 1/2, the density of the catalyst itself will be 1.6 g cm^{-3}.

Mass velocity (g h^{-1} cm^{-2})	k_g (m h^{-1})	Bulk-phase partial pressure, atm			Observed rate, (g mol SO_2 h^{-1} g cat^{-1})
		SO_2	SO_3	O_2	
251	450	0.06	.0067	0.2	0.1346
171	380	0.06	.0067	0.2	0.1278
119	340	0.06	.0067	0.2	0.1215
72	280	0.06	.0067	0.2	0.0956

a) What error prevails if, assuming linear (1st-order) kinetics in SO_2 and isothermal operation, external concentration gradients are ignored, i.e., what are the effectiveness factors?

b) Assume isothermal conditions and compute the concentration decreases to the surface, then determine the minimum and maximum external concentration gradients.

c) Assume no concentration gradient exists and compute the external temperature gradients.

d) If the reaction activation energy is 30 kcal mole^{-1}, assuming no concentration gradient, what error in rate measurement exists if an external ΔT is neglected, i.e., what are the nonisothermal effectiveness factors?

Because of the change in the Reynolds number, calculated values of k_g change; however, because of the relationship between two other dimensionless groups, the Prandtl and Schmidt numbers, assume the k_g/h ratio is constant at 9.0×10^3 cm^3 K/cal.

Problem 4.2

A commercial cumene cracking catalyst is in the form of pellets with a diameter of 0.35 cm which have a surface area, A_m, of 420 m^2 g^{-1} and a void volume, V_m, of 0.42 cm^3 g^{-1}. The pellet density is 1.14 g cm^{-3}. The measured 1st-order rate constant for this reaction at 685K was 1.49 cm^3 s^{-1} g^{-1}. Assume that Knudsen diffusion dominates and the path length is determined by the pore diameter, d_p. An average pore radius can be estimated from the relationship $\bar{r}_p = 2\bar{V}_m/A_m$ if the pores are modeled as noninterconnected cylinders (see equation 4.94). Assuming isothermal operation, calculate the Thiele modulus and determine the effectiveness factor, η, under these conditions.

Problem 4.3

A 1.0% Pd/SiO_2 catalyst for SO_2 oxidation is being studied using a stoichiometric O_2/SO_2 feed ratio at a total pressure of 2 atm. At a temperature of 673 K, a rate of 2.0 mole SO_2 s^{-1} ℓ cat^{-1} occurs. The average velocity for SO_2 molecules is 3×10^4 cm s^{-1}, and the average pore diameter in the alumina is 120 Å (10 Å = 1 nm). Assume the Pd is uniformly distributed throughout the catalyst particles. What is the largest particle size (diameter in cm) one can use in the reactor and still be assured that you have no significant diffusional effects?

Problem 4.4

Assume that Pt was dispersed throughout the pore structure of the entire pellet in Problem 4.1 and apply the Weisz-Prater criterion to determine if mass transport limitations are expected. Do only one calculation using the lowest observed rate. Assume that the average pore diameter in the catalyst is 100 Å (10 Å = 1 nm), that Knudsen diffusion dominates, and that no external transport limitations occur ($C_s = C_o$).

Problem 4.5

Vapor-phase benzene (Bz) hydrogenation over carbon-supported Pd catalysts has been studied [65,66]. A 2.1% Pd/C catalyst prepared with a carbon black cleaned in H_2 at 1223 K had a surface-weighted Pd crystallite size of 21 nm, giving a Pd dispersion of 5%, based on TEM. The carbon itself had an average mesopore diameter of 25 nm, while the average micropore diameter was 0.9 nm; thus the majority of the Pd resided in the mesopores. The highest activity of this catalyst at 413 K and 50 Torr Bz (Total P = 1 atm, balance H_2) was 1.99 μ mole Bz s^{-1} g^{-1}. The density of the catalyst was 0.60 g cm^{-3}. The catalyst particle size distribution ranged from 10–500 microns (1 μ = 10^{-6} m). (a) Assuming all the Pd is in the mesopores, are any mass transfer limitations expected based on the W-P criterion? (b) If, instead, this catalyst had all the Pd in the micropores and it gave this performance, would mass transfer limitations exist? Why?

Problem 4.6

The reforming of CH_4 with CO_2 to produce CO and H_2 has been examined over a number of dispersed metal catalysts [67,68]. At 723K, a CO_2 partial pressure of 200 Torr, $P_{CH_4} = 200$ Torr, and a total pressure of 1 atm (1 atm = 760 Torr), the catalysts listed below produced the rates listed for CO_2 consumption. The catalysts used were sieved to a 120/70 mesh fraction, thus the largest particles had a diameter of 0.020 cm. Assume a catalyst density of 1 g cm^{-3}, which provides a conservative overestimate. The pore diameter, δ, was obtained from pore size distribution measurements. Determine the W-P number for: a) Ni/SiO_2. b) Pt/ZrO_2, c) Rh/TiO_2, d) Ru/Al_2O_3. Should one be concerned about pore diffusion limitations with any of these catalysts? Why?

Catalytic Reforming of CH_4 with CO_2 (from ref. 68)		
Catalyst	$r_{CO_2}(\mu mole\ s^{-1}\ g^{-1})$	δ (nm)
Ni/SiO_2	42.6	18
Pt/ZrO_2	9.6	14
Rh/TiO_2	118	20
Ru/Al_2O_3	237	20

Problem 4.7

Utilize the data given in Problem 4.5, assume the benzene hydrogenation reaction is zero order in benzene and first order in H_2, and then calculate the Thiele modulus for the largest and smallest catalyst particles assuming all the Pd is in the mesopores. What is the Thiele modulus for the largest (500 μ) particles if all the Pd were dispersed in only the micropores? Should there be any concern about pore diffusion limitations for any of the three possibilities? Why?

Problem 4.8

The rate of formaldehyde (CH_2O) oxidation over a Ag/SiO_2 catalyst at 493 K and a CH_2O pressure of 9 Torr in air was 1.4×10^{-7} mole $CH_2O\ s^{-1}\ cm^{-3}$. The catalyst particles passed through a 100-mesh sieve (149 micron opening), and the average pore diameter of the SiO_2 was 60 Å. Can the rate be considered to be free of mass transport limitations?

5
Adsorption and Desorption Processes

An especially readable discussion of this topic is provided by Hayward and Trapnell [1]. This section will focus only on chemisorption (chemical adsorption) because physisorption (physical adsorption) is addressed in Chapter 3 as it relates to catalyst characterization.

5.1 Adsorption Rate

From the kinetic theory of gases, in a gaseous system the rate of collisions, r_{col}, between gas-phase molecules and unit surface area per unit time is proportional to the mean molecular velocity, $\bar{v}$, and the concentration of molecules, C:

$$r_{col} = \frac{\bar{v}C}{4} \tag{5.1}$$

The mean velocity of a molecule is given by $\bar{v} = \left(\frac{8k_BT}{\pi m}\right)^{1/2}$ where k_B is Boltzmann's constant, T is absolute temperature, and m is the mass of the molecule. From the ideal gas law, $PV = nRT$ or, in units of molecules, M, $PV = Mk_BT$, one obtains $C = M/V = P/k_BT$ and the collision rate can be written in terms of the pressure, P, in this system:

$$r_{col} = P/(2\pi mk_BT)^{1/2} \tag{5.2}$$

Not all of these collisions result in chemisorption, thus a sticking probability, s, is defined to represent the fraction of collisions that do provide chemisorption, thus the adsorption rate is:

$$r_{ad} = sP/\left(2\pi mk_BT\right)^{1/2} \tag{5.3}$$

The sticking probability is seldom equal to unity for any one of the following reasons.

1. Activation Energy – Chemisorption can be an activated process; therefore, only those molecules possessing the required activation energy can be chemisorbed. However, many, if not most, chemisorption processes on clean metal surfaces are nonactivated at temperatures near or above 300 K, especially for nondissociative adsorption.
2. Steric Factor – Not every molecule with the necessary activation energy will chemisorb and only those traversing the particular configuration associated with the "activated complex" will do so. Thus "s" may be much less than unity for even nonactivated chemisorption.
3. Energy Transfer Efficiency – For a gas-phase species to permanently chemisorb, it must lose a given amount of its original thermal energy during its impact with the surface, otherwise it will desorb directly. The interaction energy between a molecule and the surface is fairly high during chemisorption and collisions are frequently inelastic, which results in efficient energy transfer. This factor may be more important in physisorption, which can precede chemisorption, and thereby affect the sticking probability.
4. Surface Heterogeneity – Chemisorption capability may vary from site to site on the surface and thus change the overall observable sticking probability.
5. Occupied Sites – Collisions with occupied sites can obviously decrease chemisorption activity; however, molecules may adsorb into a weakly held second layer and migrate over occupied sites until a vacant site is found.

Different potential energy (E_{pot}) curves can be drawn to represent an adsorption process, but one of the most frequently used for physisorption is the Lennard-Jones expression:

$$E_{pot} = -ar^{-6} + br^{-12} \qquad (5.4)$$

where r is the distance from the surface, and the first term represents weak attractive forces, such as van der Waals interactions and London dispersive forces, while the latter term represents repulsive forces. Such an expression represents the potential energy of a molecule and describes the physical adsorption of a diatomic molecule, as shown in Figure 5.1. Such a representation can explain how activated adsorption can occur. The thermodynamic pathway for the dissociative chemisorption of a diatomic molecule can also be chosen such that the molecule is first dissociated to create two gas-phase atoms and these two atomic species then adsorb on the surface, a route which will almost always be nonactivated and also produce a much stronger bond, thus lowering E_{pot} to a much greater extent, as also shown in Figure 5.1. If the curves describing molecular and atomic adsorption intersect at or below the zero potential energy line, then the precursor physisorbed molecule can experience nonactivated dissociation and fall into the deep potential well for chemisorption (Figure 5.1a). In contrast, if the

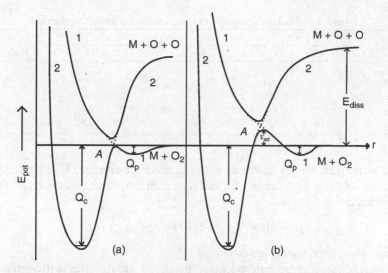

FIGURE 5.1. Potential energy curves for (1) physical and (2) chemical adsorption: Q_c, Q_p, E and E_{diss} are defined in the text. (a) Non activated, (b) Activated

energetics for these two pathways are such that the intersection occurs above the zero energy plane, then chemisorption will be activated with an activation energy, E_{ad}, as indicated in Figure 5.1b. In this figure Q_p and Q_c represent the respective heats of physisorption and chemisorption, i.e., the energy change relative to the ground state vibrational level, while E_{diss} represents the dissociation energy for the diatomic molecule. The possibility of activated adsorption was first proposed by Taylor over seventy years ago [2].

With this brief background, one can formulate a rate equation which can describe activated adsorption. For simple activated adsorption, the sticking probability can be written as:

$$s = \sigma f(\theta)e^{-E_{ad}/RT} \qquad (5.5)$$

where σ is the steric factor previously described and represents the probability that a molecule possessing sufficient energy, E_{ad}, and colliding with a vacant site will adsorb. On clean metal surfaces, initial sticking probabilities, s_o, can be quite high and approach unity, but they can also exhibit wide divergences depending on the molecule and the metal or the crystal plane on which it is adsorbing, as illustrated in Table 5.1 [3,4]. The function $f(\theta)$, where θ is the fractional surface coverage, represents the probability that a collision occurs with an empty site and the term $e^{-E_{ad}/RT}$ indicates the fraction of molecules with sufficient energy to adsorb. Consequently, from equations 5.3 and 5.5 the rate of adsorption is:

$$r_{ad} = \sigma f(\theta)e^{-E_{ad}/RT}P/(2\pi mk_BT)^{1/2} \qquad (5.6)$$

TABLE 5.1. Sticking Probabilities, s_o, on Clean Metal Surfaces

Metal	Adsorbing species									Reference
	H_2	CO	N_2	Cl_2	Br_2	I_2	O_2	O	O_3	
W(100)	0.18	0.5	0.4							3
W(110)	0.07	0.9	0.004							3
W(111)	0.23	–	0.01							3
Ge				0.25	0.3	0.3	0.02	0.4	0.3	4
Si				0.35	0.35	–	0.04	0.5	0.4	4

If the activation energy increases with increasing coverage, as is frequently observed, and if σ also varies with surface coverage, this expression can then be written as:

$$r_{ad} = \sigma(\theta)f(\theta)e^{-E_{ad}(\theta)/RT}P/(2\pi mk_BT)^{1/2} \qquad (5.7)$$

to show the dependencies on coverage.

If significant surface heterogeneity exists to create sites with different values of σ and E_ad, the surface must then be divided into a number of small elements of uniform properties, ds, and integrated over the entire surface. Various approaches exist to accomplish this and this topic will be discussed later.

5.2   Desorption Rate

In contrast to adsorption processes, which may or may not be activated as previously discussed, the desorption step is always activated, with a minimum activation energy equal to the heat of adsorption, Q_{ad}. The rate of desorption from occupied sites is:

$$r_{des} = k_{des}f'(\theta)e^{-E_{des}/RT} \qquad (5.8)$$

where k_{des} is a rate constant, $f'(\theta)$ is the fraction of covered sites available for desorption, and $E_{des} = Q_{ad} + E_{ad}$. This process is illustrated in Figure 5.2 for dioxygen, O_2.

As with adsorption, the rate constant and E_{des} can vary with coverage on a uniform surface, and the desorption rate can then be expressed as:

$$r_{des} = k_{des}(\theta)f'(\theta)e^{-E_{des}(\theta)/RT} \qquad (5.9)$$

For a simple unimolecular desorption step, an adsorbed molecule with the requisite activation energy will desorb within the period of one vibration perpendicular to the surface, hence the rate of desorption is:

$$r_{des} = \nu L\theta e^{-E_{des}/RT} \qquad (5.10)$$

where ν is the vibrational frequency and L is the site density. A comparison of Eq. 5.10 to Eq. 5.8 shows that:

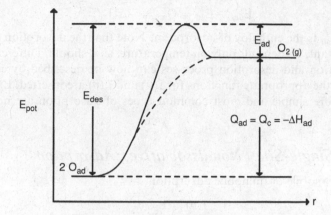

FIGURE 5.2. Potential energy changes along reaction coordinate for an associative desorption/dissociative adsorption process. The dotted curve depicts non-activated adsorption.

$$k_{des} = \nu L \qquad (5.11)$$

which is approximately equal to 10^{28} s^{-1} cm^{-2} because ν is invariably near 10^{13} s^{-1} and L is usually around 10^{15} site cm^{-2}.

5.3 Adsorption Equilibrium on Uniform (Ideal) Surfaces – Langmuir Isotherms

With this introduction, the derivation of an adsorption isotherm, which represents an equilibrium adsorption process, can be conducted [5]. The Langmuir isotherm is obtained based on the assumptions associated with an ideal uniform surface on which all sites are identical [5], i.e.,

1. Localized adsorption occurs only on vacant sites,
2. Only one adsorbed species can exist per site, that is, saturation coverage is achieved at one monolayer, and
3. The heat of adsorption is constant and independent of coverage, which assumes that no interaction occurs between adsorbed species.

At equilibrium, the rates of adsorption and desorption must be equal and, with the above assumptions, equations 5.6 and 5.8 can be used as written. Setting these two equations equal to each other and rearranging gives:

$$P = \frac{k_{des}(2\pi m k_B T)^{1/2}}{\sigma} \cdot \frac{f'(\theta)}{f(\theta)} \cdot e^{-Q_{ad}/RT} = 1/K \frac{f'(\theta)}{f(\theta)} \qquad (5.12)$$

because Figure 5.2 shows that:

$$E_{des} - E_{ad} = Q_{ad} = -\Delta H_{ad}, \tag{5.13}$$

where ΔH_{ad} is the enthalpy of adsorption. Note that the adsorption equilibrium constant, K, depends only on temperature, as it should. Different types of adsorption and desorption processes can now be described by equation 5.12 after the appropriate functions for $f(\theta)$ and $f'(\theta)$ are inserted. Examples for the more simple and most common types of adsorption are now provided.

5.3.1   Single-Site (Nondissociative) Adsorption

For either mobile or immobile adsorption

$$f(\theta) = 1 - \theta \tag{5.14}$$

and

$$f'(\theta) = \theta \tag{5.15}$$

Substitution into equation 5.12 gives

$$P = \theta/K(1 - \theta) \tag{5.16}$$

and the Langmuir isotherm relating coverage to pressure is then obtained:

$$\theta = n/n_m = KP/(1 + KP) = K'[A]/(1 + K'[A]) \tag{5.17}$$

where n is the adsorbate uptake (mole g^{-1}) at a given pressure and n_m is the maximum uptake, with the assumption that the latter corresponds to saturation or monolayer coverage, and [A] is the concentration of an ideal gas, A. Instead of n/n_m, most older studies and some recent ones use v/v_m where v is the volume adsorbed at a given pressure and v_m is the monolayer coverage, *with both values always reported at STP conditions*. Such an approach is needed experimentally because θ is typically not an observable variable. The surface concentration of an adsorbed species, A, is then

$$[A]_s = C_{A_s}(\text{molecule area}^{-1}) = L\theta \tag{5.18}$$

5.3.2   Dual-Site (Dissociative) Adsorption

Both mobile and immobile coverages can again be considered but now, because the desorption step involves the recombination of two neighboring adsorbed species, the functions of $f(\theta)$ and $f'(\theta)$ are different than before because of the difference in reaction (recombination) probabilities.

a) Mobile Adsorbed Species – This is the simpler case and it depends only on the surface concentration hence, for a diatomic molecule based on the law of mass action:

$$f(\theta) = (1 - \theta)^2 \tag{5.19}$$

and

$$f'(\theta) = \theta^2 \qquad (5.20)$$

which gives

$$P = \theta^2/K(1-\theta)^2 \qquad (5.21)$$

or

$$\theta = K^{1/2}P^{1/2}/(1 + K^{1/2}P^{1/2}) \qquad (5.22)$$

b) Immobile Adsorbed Species – Dissociation of a diatomic molecule into two atoms which cannot move off their sites results in the requirement that site pairs are needed, then

$$f(\theta) = \left(\frac{Z}{Z-\theta}\right)(1-\theta)^2 \qquad (5.23)$$

and

$$f'(\theta) = \left(\frac{(Z-1)^2}{Z(Z-\theta)}\right)\theta^2 \qquad (5.24)$$

where Z is the number of nearest neighbor sites [1], and the isotherm becomes

$$\theta = \left(\frac{Z}{Z-1}\right)K^{1/2}P^{1/2}/\left(1 + \left(\frac{Z}{Z-1}\right)K^{1/2}P^{1/2}\right) = (K'P)^{1/2}/\left(1 + (K'P)^{1/2}\right)$$

$$(5.25)$$

where

$$K' = \left(\frac{Z}{Z-1}\right)^2 K. \qquad (5.26)$$

Thus these two systems give identical mathematical forms for the Langmuir equation and they cannot be differentiated experimentally by uptake measurements alone. However, for immobile adsorption requiring site pairs, it has been shown statistically that complete saturation cannot be achieved and the highest coverage is about 90% of a full monolayer [6].

These familiar forms of the Langmuir isotherm describing nondissociative and dissociative adsorption are illustrated in Figure 5.3. Saturation occurs in the monolayer region at high pressures, while the 1st-order dependence that exists at low pressures is referred to as the Henry's law region, i.e., $\theta_A = K_A P_A$.

One straightforward way to test these isotherms is to linearize them, evaluate the linearity and, if acceptable, obtain values for the monolayer coverage, n_m, and the adsorption equilibrium constant, K, from the slope and the intercept. For example, equation 5.17 for the single-site model can be rearranged to:

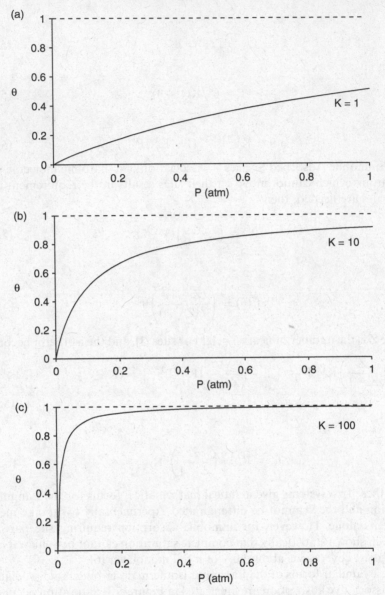

FIGURE 5.3. Plots of coverage θ versus pressure P for a Langmuir isotherm representing nondissociative adsorption on a single site: (a) low value of K_A; (b) intermediate value of K_A; (c) high value of K_A.

$$\frac{P}{n} = \frac{1}{Kn_m} + \frac{P}{n_m} \tag{5.27}$$

while for dual-site adsorption, linearization of either equation 5.22 or 5.25 gives

$$\frac{P^{1/2}}{n} = \frac{1}{K^{1/2}n_m} + \frac{P^{1/2}}{n_m}$$

(5.28)

In either of these arrangements, the slope gives the monolayer uptake.

5.3.3 Derivation of the Langmuir Isotherm by Other Approaches

The Langmuir isotherm can also be derived by other methods including statistical mechanics, thermodynamics, and chemical reaction equilibrium. The last approach is especially straightforward and useful, and it is developed as follows. For nondissociative chemisorption, the adsorption step is represented as a reaction, i.e., for an adsorbing gas-phase molecule, A, which adsorbs on a site, *:

(5.1)
$$A_{(g)} + * \;\overset{K_A}{\rightleftharpoons}\; A*$$

thus from thermodynamics, where K_A is an adsorption equilibrium constant:

$$K_A = C_{A*}/P_A C_v$$

(5.29)

where C_{A*} and C_v are surface concentrations (per unit area) of adsorbed species and vacant sites, respectively, which can be represented by $L\theta_A$ and $L\theta_v$. A site balance can always be used to eliminate one unknown, C_v, in this case, where L is the total site density (or site concentration):

$$L = C_{A*} + C_v$$

(5.30)

or, alternatively from Eq. 5.18:

$$1 = \theta_A + \theta_V$$

(5.31)

$$\text{then } C_V/L = \theta_V = 1 - \theta_A$$

(5.32)

$$\text{and } K_A = L\theta_A/P_A L(1 - \theta_A)$$

(5.33)

which, after rearrangement, gives:

$$\theta_A = K_A P_A/(1 + K_A P_A)$$

(5.34)

which is identical to equation 5.17.

In similar fashion, the dissociative adsorption of a diatomic gaseous molecule A_2 is written:

(5.2)
$$A_{2(g)} + 2* \;\overset{K_{A_2}}{\rightleftharpoons}\; 2A*$$

and

$$K_{A_2} = C_{A*}^2/P_{A_2} C_v^2$$

(5.35)

A site balance again gives equation 5.30 (or 5.31) and substitution into equation 5.35 leaves

$$K_{A_2} = \theta_A^2 / P_{A_2}(1 - \theta_A)^2 \tag{5.36}$$

Taking the square root of both sides of the equation and rearranging gives:

$$\theta_A = K_{A_2}^{1/2} P_{A_2}^{1/2} \Big/ \left(1 + K_{A_2}^{1/2} P_{A_2}^{1/2}\right) \tag{5.37}$$

which is identical to equations 5.22 (and 5.25).

The K's in these Langmuir isotherms represent adsorption equilibrium constants, thus they can be represented in the following ways:

$$K = e^{-\Delta G_{ad}^o/RT} = e^{\Delta S_{ad}^o/R} e^{-\Delta H_{ad}^o/RT} = K_o e^{Q_{ad}/RT}, \tag{5.38}$$

where ΔG_{ad}^o, ΔH_{ad}^o, and ΔS_{ad}^o are the changes in the standard Gibbs free energy, enthalpy, and entropy, respectively, for the adsorption processes represented by steps 5.1 and 5.2. Changes in thermodynamic properties are typically referenced to a standard state, and it is of some interest to determine this standard state for Langmuirian adsorption. As is routinely done, the standard state for a gas is defined as that where its activity, a, equals its partial pressure, P_o, which is equal to 1 atmosphere, i.e.,

$$a = P_o = 1 \text{ atm} \tag{5.39}$$

For any reaction at equilibrium:

$$RT \ln K = -\Delta G^o \tag{5.40}$$

and for an equilibrium adsorption process, as indicated by equation 5.38:

$$\Delta G_{ad}^o = \Delta H_{ad}^o - T\Delta S_{ad}^o \tag{5.41}$$

The thermodynamic pathway shown in Figure 5.4 can be proposed to describe the adsorption of a gas at equilibrium from the given standard state:
For step 1 the change in standard free energy is:

$$\Delta G_1^o = RT \ln P - RT \ln P_o = RT \ln P \tag{5.42}$$

For step 2, $\Delta G_2^o = 0$ because this step is at equilibrium, therefore, for the overall adsorption process,

$$\Delta G_{ad}^o = \Delta G_1^o + \Delta G_2^o = RT \ln P = -RT \ln K. \tag{5.43}$$

Consequently, $K = P^{-1}$ and the standard state for adsorption corresponds to:

$$\theta_{ss} = KP/(1 + KP) = 1/(1 + 1) = 1/2 \tag{5.44}$$

This value of θ results in a configurational entropy term equal to zero.

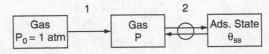

FIGURE 5.4. Thermodynamic representation of the standard state (SS) for an adsorption process.

5.3.4   Competitive Adsorption

The adsorption systems discussed to this point have involved only one adsorbate; however, the effect of additional adsorbates creating competitive adsorption for sites can be readily accommodated by using the site balance. For example, assume that two adsorbates, A and B, adsorb simultaneously and compete for the same sites, i.e.,

$$A_{(g)} + * \overset{K_A}{\rightleftharpoons} A* \tag{5.3}$$

and

$$B_{(g)} + * \overset{K_B}{\rightleftharpoons} B* \tag{5.4}$$

The site balance gives:

$$L = C_v + C_{A*} + C_{B*} \tag{5.45}$$

or, because $C_{A*}/L = \theta_A$, $C_{B*}/L = \theta_B$ and $C_v/L = \theta_v$:

$$1 = \theta_v + \theta_A + \theta_B \tag{5.46}$$

then

$$\theta_v = 1 - \theta_A - \theta_B \tag{5.47}$$

and

$$K_A = \theta_A/P_A\theta_v = \theta_A/P_A(1 - \theta_A - \theta_B) \tag{5.48}$$

while

$$K_B = \theta_B/P_B\theta_v = \theta_B/P_B(1 - \theta_A - \theta_B) \tag{5.49}$$

Solving these two equations with two unknowns produces:

$$\theta_A = K_A P_A/(1 + K_A P_A + K_B P_B) \tag{5.50}$$

and

$$\theta_B = K_B P_B/(1 + K_A P_A + K_B P_B) \tag{5.51}$$

Thus, in general for nondissociative adsorption, the fractional coverage of a particular species, such as A, can readily be shown to be:

$$\theta_A = K_A P_A/(1 + \Sigma K_i P_i) \tag{5.52}$$

which includes any surface species whose adsorption is equilibrated, such as products, inhibitors and other nonreacting species. The fraction of the surface remaining bare is:

$$\theta_v = 1/(1 + \Sigma K_i P_i) \tag{5.53}$$

Application of this approach to dissociative adsorption of a two-component system, A_2 and B_2, gives the same site balance as equation 5.46, and utilization of equation 5.48 for each species results in:

$$\theta_A = K_A^{1/2}P_A^{1/2} \Big/ \left(1 + K_A^{1/2}P_A^{1/2} + K_B^{1/2}P_B^{1/2}\right) \tag{5.54}$$

$$\text{and} \quad \theta_B = K_B^{1/2}P_B^{1/2} \Big/ \left(1 + K_A^{1/2}P_A^{1/2} + K_B^{1/2}P_B^{1/2}\right) \tag{5.55}$$

One can, of course, have systems where both nondissociative and dissociative adsorption occurs simultaneously, and this is all easily handled in the site balance. Consequently, the summation term in equations 5.52 and 5.53 would be expanded to $\sum K_i P_i + \sum K_j^{1/2}P_j^{1/2}$ to include both nondissociative and dissociative adsorption respectively. However, if nondissociative adsorption involving two or more sites is proposed, such as

(5.5)
$$A_{(g)} + 2* \overset{K_A}{\rightleftarrows} *A* $$

for example, then the site balance becomes $1 = \theta_v + 2\theta_A$, or $L = C_v + 2C_{*A*}$, and the solution to these two equations with two unknowns to get the coverage of A necessitates the use of the quadratic formula and the solution is much more complex. Furthermore, one could use equation 5.52 to differentiate between an inhibitor and a poison, two terms that are frequently used interchangeably. The negative influence of an inhibitor should be dependent upon its partial pressure in the system and should exhibit a reversible effect. In contrast, a poison could be considered to be so strongly bound that its effect is irreversible, thus decreasing its (or its precursor's) partial pressure or concentration to zero does not reduce its surface concentration.

5.4   Adsorption Equilibrium on Nonuniform (Nonideal) Surfaces

5.4.1   The Freundlich Isotherm

It is, of course, unrealistic to expect real surfaces to always behave as the ideal surface defined by the previous assumptions, and many adsorption studies have reported data that do not fit a Langmuir isotherm. Among the early efforts to correlate such data was the empirical isotherm proposed by Freundlich in 1926 [7], i.e.,

$$n = cP^{1/a}\left(\text{or } \theta = n/n_m = c'P^{1/a}\right) \tag{5.56}$$

where c and a are dependent on T, c also depends on the surface area of the adsorbent, and a is always greater than unity. Usually, both c and a decrease with increasing temperature. Because the amount adsorbed is proportional to a fractional power of the pressure, the Freundlich isotherm is similar to the Langmuir isotherm in regions of moderate coverage and, in fact, the two isotherms can be almost coincident over a wide range of pressure by the appropriate choices of the two constants [1]. Subsequent to its proposal, the Freundlich equation has been derived by both a thermodynamic approach and a statistical approach [8] in which a heterogeneous surface is divided into small domains containing identical sites and the total coverage is the sum of the coverages within each domain, as originally suggested by Langmuir [5]. Thus, for single-site adsorption within each domain:

$$\theta_i = K_i P/(1 + K_i P) \tag{5.57}$$

and the overall coverage is

$$\theta = \Sigma n_i \theta_i \tag{5.58}$$

where n_i is the fraction of type i sites. If the K values for each domain are assumed to be close enough in value so that a continuous distribution can be assumed, then equation 5.58 becomes:

$$\theta = \int n_i \theta_i di \tag{5.59}$$

where $n_i di$ is the frequency of occurrence of θ_i between i and i + di. Integration of this expression cannot be accomplished without a distribution function to substitute for n_i; however, if it is assumed that i is represented by the heat of adsorption, Q_{ad}, and that n_i is exponentially dependent on Q_{ad}, then:

$$n_{Q_{ad}} = n_o e^{-Q_{ad}/Q_{ad_m}} \tag{5.60}$$

where n_o and Q_{ad_m} are constants, then a Freundlich isotherm can be obtained after integration [8]. A subsequent approach by Halsey and Taylor is quite understandable and is as follows [9], keeping in mind that Q_{ad} now represents the property i. Equations 5.57 and 5.60 are substituted into equation 5.59 to give:

$$\theta = \int_0^\infty n_{Q_{ad}} \left(\frac{K_{Q_{ad}} P}{(1 + K_{Q_{ad}} P)} \right) dQ_{ad} \tag{5.61}$$

Remember from equation 5.38 that $K = K_o e^{Q_{ad}/RT}$, where K_o is a constant. Substituting this into equation 5.61 and rearranging gives:

$$\theta = \int_0^\infty \frac{n_{Q_{ad}} dQ_{ad}}{\left(1 + \frac{1}{K_{Q_{ad}} P} \right)} = \int_0^\infty \frac{n_o e^{-Q_{ad}/Q_{ad_m}} dQ_{ad}}{1 + \frac{e^{-Q_{ad}/RT}}{K_o P}} \tag{5.62}$$

If one postulates that $Q_{ad} \gg \pi RT$, which is satisfied once Q_{ad} exceeds about three times the thermal energy, which is reasonable (this value is less than 1.9 kcal mole^{-1} at room temperature), equation 5.62 can be integrated to give:

$$\theta = n_o Q_{ad_m} (K_o P)^{RT/Q_{ad_m}} \tag{5.63}$$

It can be shown that when $\theta = n_o Q_{ad_m}$, this corresponds to maximum adsorption, i.e., $\theta = 1$, so this equation can be normalized to $(K_o P)^{RT/Q_{ad_m}}$, which is the form of the empirical Freundlich isotherm given by equation 5.56.

5.4.2 The Temkin Isotherm

One might anticipate that the Freundlich isotherm would be more widely applicable for experimental results because it allows for a decrease in the heat of adsorption, as opposed to a Langmuir isotherm. However, such an exponential decrease in Q_{ad} is not often found, whereas a linear or near-linear decrease is observed much more often. An excellent example of this is the variation in Q_{ad} for CO chemisorbed on different crystal planes of Pt, as reported by McCabe and Schmidt [10] and shown in Table 5.2. The Temkin isotherm is predicated on this assumption [11] and can be obtained by substituting a linear dependence of Q_{ad} on θ into the adsorption equilibrium constant contained in a Langmuir isotherm:

$$K = K_o e^{Q_{ad_o}(1-\alpha\theta)/RT} \qquad (5.64)$$

where Q_{ad_o} is the initial heat of adsorption on a clean surface and α is a constant. Stipulating now a uniform surface with all sites equal and choosing single-site adsorption as an example, equation 5.16 can be reorganized to:

$$PK = PK_o e^{Q_{ad_o}(1-\alpha\theta)/RT} = \theta/(1-\theta) \qquad (5.65)$$

Logarithms of each side are taken and the equation is transposed to:

$$\ln P = -\ln\left(K_o e^{Q_{ad_o}/RT}\right) + \alpha Q_{ad_o}\theta/RT + \ln\theta/(1-\theta) \qquad (5.66)$$

Note that $K_o e^{Q_{ad_o}/RT}$ represents a term, B_o, which is independent of θ, and at intermediate values of θ, the last term varies relatively little with θ (± 1.4 between $\theta = 0.2$ and 0.8). For chemisorption $\alpha Q_{ad_o} \gg RT$, thus $\ln P$ is primarily dependent on $\alpha Q_{ad_o}\theta/RT$. Therefore, in regions of θ not approaching zero or unity, to a good approximation equation 5.66 becomes the Temkin equation, which gives a linear relationship between θ and $\ln P$, i.e.,

$$\theta = \frac{RT}{\alpha Q_{ad_o}} \ln(B_o P) = A_o \ln(B_o P) \qquad (5.67)$$

If a nonuniform surface is now considered, it is again necessary to divide the surface into domains, ds, each of which has uniform sites giving a constant heat of adsorption [1]. Within each of these domains, a Langmuir isotherm is

TABLE 5.2. Variation of the Heat of Adsorption, Q_{ad}, with Coverage $[Q_{ad} = Q_{ad_o}(1-\alpha\theta)]$ and Sticking Probabilities (s_o) for CO on Different Pt Crystal Planes (from Ref. 10)

Pt crystal plane	T(K)	Q_{ad_o}(kcal/mole)	α	s_o @ 300K
(111)	535	29.6	6.5	0.34
(110)	460	26.0	2.5	0.64
(100)	550	31.9	<1.0	0.24
(200)	625	36.2	3.0	0.95
(211)	610	35.2	6.5	0.27

assumed applicable and the value of θ for the entire surface is obtained by integration:

$$\theta = \int \theta_s ds \tag{5.68}$$

If now a linear decrease in Q_{ad} with coverage is again assumed and equation 5.64 is used then, with the total surface area, S, being normalized to unity, the equation to integrate is:

$$\theta = \int_0^1 \frac{B_o Pe^{-\alpha Q_{ad_o} s/RT}}{1 + B_o Pe^{-\alpha Q_{ad_o} s/RT}} ds \tag{5.69}$$

The integral gives

$$\theta = \frac{RT}{\alpha Q_{ad_o}} \ln\left(\frac{1 + B_o P}{1 + B_o Pe^{-\alpha Q_{ad_o}/RT}}\right) \tag{5.70}$$

In the middle range of coverage, where the pressure is assumed high enough so that $B_o P \gg 1$, but still low enough so that $B_o Pe^{-\alpha Q_{ad_o}/RT} \ll 1$, this equation simplifies to equation 5.67. If dissociative adsorption is assumed, the derived equation is still identical to equation 5.67 [1].

To test these various isotherms to determine which best fits the experimental data, it is most convenient to linearize them. Thus for the different isotherms one gets:

Langmuir isotherm, single-site: $P/n = P/n_m + 1/(Kn_m)$ $\qquad$ (5.71)

Langmuir isotherm, dual-site: $P^{1/2}/n = P^{1/2}/n_m + 1/\left(K^{1/2}n_m\right)$ $\quad$ (5.72)

Freundlich isotherm: $\ln n = \ln n_m K_o + (RT/Q_{ad_m})\ln P$ $\qquad$ (5.73)

Temkin isotherm: $n = n_m(RT/\alpha Q_{ad_o})(\ln B_o + \ln P)$ $\qquad$ (5.74)

The dependencies of these three types of isotherms on the heat of adsorption is illustrated in Figure 5.5. Some adsorbate/adsorbent systems approximate Langmuir behavior up to certain levels of coverage [12,13], whereas many more exhibit Temkin behavior, as shown in Table 5.2.

5.5 Activated Adsorption

It is worth mentioning a commonly observed chemisorption process, i.e., a relatively slow uptake of adsorbate following a rapid initial uptake, which is described as activated adsorption. The rate of uptake frequently obeys the relationship

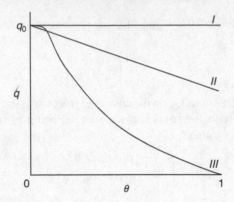

FIGURE 5.5. Variation of the heat of adsorption q (Q_{ad} in the text) with coverage for various isotherms: I – Langmuir isotherm, II – Temkin isotherm, III – Freundlich isotherm. (Reprinted from ref. 1, copyright © 1964, with permission from Elsevier)

$$r_{ad} = \frac{dq}{dt} = \frac{dN_i}{Adt} = ae^{-bq}, \qquad (5.75)$$

where q is the amount adsorbed (mole of species i per unit surface area) and a and b are constants. Equation 5.75 is known as the Elovich equation [14], and it can be derived for either a uniform or a nonuniform surface by assuming that the activation energy for adsorption varies with q [1]. For example, for simplicity, consider single site adsorption on a uniform surface where $f(\theta) = (1 - \theta)$ and E_{ad} varies linearly with coverage according to the relationship:

$$E_{ad} = E_o + \alpha\theta \qquad (5.76)$$

where E_o and α are constants. Substituting the above values into equation 5.6 yields

$$r_{ad} = \frac{\sigma P}{(2\pi mk_B T)^{1/2}} (1 - \theta)e^{-(E_o + \alpha\theta)/RT} \qquad (5.77)$$

Now, if σ is assumed to be relatively invariant and θ is restricted to be not near unity so that the variation in $(1 - \theta)$ can be neglected, then, writing the rate in terms of fractional coverage:

$$\frac{r_{ad}}{L} = \frac{d\theta}{dt} = k_{ad}e^{-\alpha\theta/RT} \qquad (5.78)$$

which has the form of the Elovich equation. The same form can also be derived if the total number of surface sites is not assumed constant, but is dependent on both q and T [1]. Integration of equation 5.78 gives

$$\theta = (RT/\alpha)\ln\left(\frac{t + t_o}{t_o}\right) \qquad (5.79)$$

where $t_o = RT/\alpha k_{ad}$ and k_{ad} has a different T dependence for uniform and nonuniform surfaces [1]. Consequently, a plot of θ (or q) vs. $\ln(t + t_o)$ should be linear.

Although this relationship is often obeyed, in many cases it is not and the slope changes abruptly [1]. Each different slope has been interpreted to represent a different set of adsorption sites with its own values for E_o and α. However, an alternative perspective of time-dependent adsorption was proposed more recently by Ritchie [15], which in many cases removed the non-linearity observed in Elovich plots. His approach did not necessitate the assumption of a variable activation energy, and he simply assumed that more than one site may be required in the adsorption process. Therefore, the rate of change of surface coverage could be written as:

$$r_{ad} = \frac{d\theta}{dt} = k_{ad}(1 - \theta)^n \tag{5.80}$$

where k_{ad} is a rate constant and n is the number of sites occupied by each molecule of adsorbate. If $n = 1$, equation 5.80 integrates to

$$\theta = \frac{N_{ad}}{N_\infty} = 1 - e^{-k_{ad}t} \tag{5.81}$$

or alternatively

$$\ln\left(\frac{N_\infty - N_{ad}}{N_\infty}\right) = -k_{ad}t, \tag{5.82}$$

and if $n > 1$, equation 5.80 integrates to

$$\frac{N_\infty^{n-1}}{(N_\infty - N_{ad})^{n-1}} = (n - 1)k_{ad}t + 1 \tag{5.83}$$

where N_{ad} is the amount adsorbed at any time, t, and N_∞ is the amount adsorbed at infinite time, i.e., the maximum coverage. If $n = 2$, for example, which is a very likely possibility, one would obtain

$$\frac{N_\infty}{(N_\infty - N_{ad})} = k_{ad}t + 1, \tag{5.84}$$

and this choice linearized earlier non-linear Elovich plots for H_2 adsorption on Graphon and Mo/Al_2O_3 and for H_2O on Vycor [15]. Thus one should also want to consider this model when correlating uptake versus time.

References

1. D. O. Hayward and B. M. W. Trapnell, "Chemisorption", Buttersworth, London, 1964.
2. H. S. Taylor, *J. Am. Chem. Soc.* 53 (1931) 578.
3. L. D. Schmidt, *Catal. Rev.* 9 (1974) 115.

4. R. J. Madix and A. Susu, *J. Catal.* 28 (1973) 316.
5. I. Langmuir, *J. Am. Chem. Soc.* 40 (1918) 1361.
6. A. R. Miller, *Proc. Cambridge Phil. Soc.* 43 (1947) 232.
7. H. Freundlich, "Colloid and Capillary Chemistry", Methuen, London, 1926.
8. J. Zeldowitsch, *Acta Physicochim.* U.R.S.S. 1 (1934) 961.
9. G. Halsey and H. S. Taylor, *J. Chem. Phys.* 15 (1947) 624.
10. R. W. McCabe and L. D. Schmidt, *Surf. Sci.* 66 (1977) 101.
11. A. Frumkin and A. Slygin, *Acta Physicochim.* U.R.S.S. 3 (1935) 791.
12. K. Christmann, G. Ertl, and T. Pignet, *Surf. Sci.* 54 (1976) 365.
13. N. Vasquez, A. Muscat and R. J. Madix, *Surf. Sci.* 301 (1994) 83.
14. S. Yu. Elovich and G. M. Zhabrova, *Zh. Fiz. Khim.* 13 (1939) 1761.
15. A. G. Ritchie, *J. Chem. Soc.* Faraday Trans. I. 73 (1977) 1050.
16. J. Shen and J. M. Smith, *Ind. & Eng. Chem. Fund.* 7 (1968) 100.
17. K. J. Laidler, "Chemical Kinetics", Harper & Row, NY, 1987.

Problem 5.1

Derive the form of the Langmuir isotherm for the complete dissociative adsorption of SO_2, i.e.,

$$SO_2 + 3* \overset{K}{\rightleftharpoons} 2O - * + S - *$$

where * is an empty site. Write the isotherm in terms of θ, the fraction of sites covered by any atom. What is the expression for the fraction of sites covered by sulfur atoms, θ_s?

Problem 5.2

The following data have been reported by Shen and Smith for benzene (Bz) adsorption on silica gel [16]:

P_{Bz} (atm)	μmol Bz adsorbed/g SiO_2			
	343 K	363 K	383 K	403 K
1.0×10^{-3}	220	112	45	20
2.0×10^{-3}	340	180	78	39
5.0×10^{-3}	680	330	170	86
1.0×10^{-2}	880	510	270	160
2.0×10^{-2}	—	780	420	260

a) Do these data better fit a single site or a dual site Langmuir isotherm? Why?

b) Assuming single site adsorption, calculate the enthalpy and entropy of adsorption for benzene on SiO_2

c) What is saturation coverage at each temperature?

Problem 5.3

Take the data at 363 K from problem 5.2. Can you distinguish which of the three isotherms – Langmuir, Freundlich, or Temkin – is the best obeyed? If so, which one?

Problem 5.4

Derive the form of the Langmuir isotherm when an adsorbing species occupies two surface sites, i.e.,

$$A + 2S \overset{K}{\rightleftharpoons} \text{S-A-S}$$

6
Kinetic Data Analysis and Evaluation of Model Parameters for Uniform (Ideal) Surfaces

Obviously, one should work with kinetic data free from any significant artifacts, particularly heat and mass transfer limitations, and methods to verify the absence of the latter complications have been discussed in Chapter 4. In addition, the equilibrium conversion for the reaction of interest must be determined, either directly from thermodynamic tables of free energies or, if not available, from estimates of free energies of formation based on group properties [1]. The first goal of any kinetic study is to get an accurate predictive relationship between the rate of reaction and the experimental parameters controlling it, which are primarily the temperature and the concentrations (or partial pressures) of the reactants. If conversions are high enough, then consideration of the influence of the products may also be required, both in regard to competitive adsorption and to the reverse reaction if conversions are too close to the equilibrium conversion. As mentioned previously, rate data acquired under differential reactor conditions are easier to interpret, not only because the reactant concentrations can be assumed to be essentially constant, but also because the product concentrations are very low and their influence can typically be ignored. To put these data into a predictive framework, it is invariably beneficial to strive for a second goal, which is to obtain a reaction model based on a series of elementary steps. As defined earlier, an elementary step must be written exactly as it occurs on a molecular level. Such a model provides insight into the surface chemistry and allows it to be tested by various methods.

In all heterogeneous catalytic reactions, the initial step must involve the adsorption of at least one of the reactants from the gas or liquid phase. This step is then followed by a series of elementary steps that describe the reaction occurring on the surface, and the final steps represent desorption of the products. Consequently, from an overall perspective a general representation of a catalytic process involves five consecutive steps:

1. Transport of the reactants to the catalytic surface.
2. Adsorption of the reactants on the active sites.
3. Reaction on the surface.
4. Desorption of the products from the active sites.
5. Transport of the products into the surrounding phase.

If one works with data collected in the kinetic regime, i.e., free from mass transfer limitations, then steps 1 and 5 are very rapid on a relative scale and can be ignored. Thus the challenge is to acquire data in the kinetic regime and then describe the surface reaction at a microscopic level by a series of elementary steps which are used to derive a rate expression relating the unknown concentrations of surface intermediates to observable macroscopic concentrations of reactants (and products, if necessary).

The simplest method to obtain such a rate equation involves the use of a Langmuir isotherm to relate the coverage, or surface concentration, of a reactant or product species with its gas-phase or liquid-phase concentration, which can be measured. This is an approach originally employed by Langmuir [2] and subsequently used frequently by Hinshelwood [3,4], and the resulting rate expressions are referred to as Langmuir-Hinshelwood models. Implicit in this model is the assumption that all adsorption/desorption steps involving reactants and products are quasi-equilibrated and either the rate determining step (RDS) or the slow steps if a RDS does not exist involve reactions on the catalyst surface. Also included in this model are all the assumptions associated with the derivation of a Langmuir isotherm, but they pertain just to the sites that dominate the rate in the catalytic sequence, that is, the site density L contained in the rate constant k may not correspond on a one-to-one basis with the sites measured by adsorption under non-reacting conditions. As just suggested, there can be many instances where a RDS does not exist. What approach is used then? Let us consider the possibilities.

In modeling reactions, in general, and catalytic reactions, in particular, the kineticist must draw on as many tools at his disposal as possible. Some of the most important concepts that are routinely used to derive, simplify and evaluate complicated rate expressions are: 1) Transition-state theory; 2) The steady-state approximation; 3) Bond-order conservation calculations for surface species; 4) A rate determining step; 5) A most abundant reaction intermediate; and 6) Criteria to evaluate parameters in derived rate expressions. Let us examine these topics prior to their utilization in deriving and evaluating reaction models and rate equations.

6.1   Transition-State Theory (TST) or Absolute Rate Theory

A brief background on this subject will prove beneficial when studying the kinetics of catalytic reactions. The conventional transition-state theory (TST) of reaction rates was published separately during the same year

(1935) by Eyring [5] and by Evans and Polanyi [6], and this theory is also referred to as "absolute rate theory" or "activated-complex theory". A very readable description of this topic is provided by Laidler [7], and the following discussion of the derivation of this theory is drawn heavily from this reference. Improvements and modifications have been made to the original theory [7], but an examination of only the conventional TST will suffice for our purposes.

The TST rate equation can be derived in various ways, and one of the most straightforward approaches which utilizes the quasi-equilibrium hypothesis will be chosen here. As an example, consider the plot of potential energy vs. the minimum-energy path representing the reaction coordinate for an exothermic gas-phase reaction, $A + B \rightleftharpoons C$, as shown in Figure 6.1.

One can view the formation of the activated complex, $X^{\ddagger}$, as being quasi-equilibrated while its decomposition into the products is an irreversible rate determining step; consequently, for the reaction $A + B \Rightarrow C$, this process can be represented as:

$$
(6.1) \qquad A + B \underset{k_{-1}}{\overset{k_1}{\rightleftharpoons}} X^{\ddagger}
$$

$$
(6.2) \qquad X^{\ddagger} \xrightarrow{k_2} C
$$

where, by definition of a RDS, k_1 and $k_{-1} \gg k_2$. Thus from equilibrium thermodynamics:

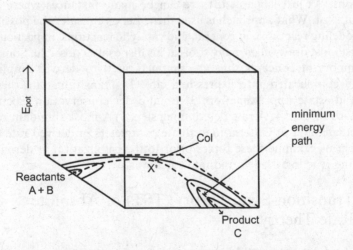

FIGURE 6.1. Potential energy surface within a given region showing the col, or saddle point, and the minimum-energy path, which is that of steepest descent in either direction from the col. The activated complex $X^{\ddagger}$ exists at the col.

$$\frac{k_1}{k_{-1}} = K^{\ddagger} = \frac{C_{X^{\ddagger}}}{C_A C_B} = \frac{[X^{\ddagger}]}{[A][B]} = \frac{Q_{X^{\ddagger}}}{Q_A Q_B} e^{-E_o/RT} \qquad (6.1)$$

where either C_i or [i] represents the concentration of species i. The equilibrium constant for the first step can also be expressed in terms of statistical thermodynamics whereby Q_i represents the partition function for species i (per unit volume), and E_o is the energy increase per mole at absolute zero for the formation of 1 mole $X^{\ddagger}$; thus E_o represents the activation energy at zero degrees K for the reaction given by step 6.1 because all components are in their ground states.

A partition function is the summation of all the permissible energy levels determined from quantum mechanics for a particular system, and the total partition function for a molecular or atomic species is defined by:

$$Q \equiv \sum_i g_i e^{-\varepsilon_i/k_B T} \qquad (6.2)$$

where g_i is the degeneracy, i.e., the number of energy states allowed in the ith level, k_B is Boltzmann's constant, and ε_i is the energy of the ith state relative to the zero-point energy. Equation 6.2 relates to the probability, $P(\varepsilon_i)$, that a molecule will be in a certain energy state, ε_i, because the Boltzmann law states the proportionality:

$$P(\varepsilon_i) \, \alpha \, g_i e^{\varepsilon_i/k_B T} \qquad (6.3)$$

which impacts on the energy distribution. It is usually assumed that the different forms of energy – translational, rotational, vibrational and electronic – are independent of each other. If so, the total energy corresponding to the ith state can then be expressed as the sum of the four respective different forms of energy, i.e.,

$$\varepsilon_i = t_i + r_i + v_i + e_i \qquad (6.4)$$

and the partition function can be written as

$$Q = q_t q_r q_v q_e = \sum_i g_{ti} e^{-t_i/k_B T} g_{ri} e^{-r_i/k_B T} g_{vi} e^{-v_i/k_B T} g_{ei} e^{-e_i/k_B T} \qquad (6.5)$$

With this factorization of the total partition function, each term can be determined separately.

For the electronic states for a species, one must know the electronic energy levels, e_i, and the electronic partition function is then:

$$q_e = \sum_i g_{ei} e^{-e_i/k_B T} \qquad (6.6)$$

However, at typical temperatures this term seldom makes a significant contribution to the partition function because the excited electronic energy levels are usually too high to be populated, and only if these levels are less than $4 k_B T$ do they begin to be significant [7]. If the lowest energy state is nondegenerate, as with most molecules, then $g_{ei} = 1$ and $q_e \cong 1$; but for a few molecules, such as O_2 and NO, the lowest level is not a singlet and q_e is no longer unity. For the other forms of energy, expressions have been derived for the individual partition functions [8], and they are listed in Table 6.1. Order of magnitude estimates of each are also provided. The partition function for translational motion involves the mass of the species. The partition function for rotational motion includes the moment of inertia, I, which can vary by a factor of 2 for a symmetrical linear molecule compared to a nonlinear molecule.

The total number of degrees of freedom in a molecule containing N atoms is 3N. For a molecule in a fluid phase, such as a gas, three degrees are associated with translational motion, three degrees are associated with rotational energy for a non-linear molecule, but only two degrees exist for a linear molecule, thus the remaining degrees of freedom, 3N-6 or 3N-5, respectively, must relate to vibrational motion. The expression given for the latter in Table 6.1 is for a single degree of vibrational freedom, i.e., a normal mode of vibration.

With this brief introduction to partition functions, let us return to Figure 6.1 and derive a rate of reaction, as represented by step 6.2, which can be thought of as a very loose vibration resulting in bond rupture (or formation) and passage over the peak of the energy pathway to give the product. Equation 6.1 describes the quasi-equilibrated formation of the activated complex, which possesses either $3(N_A + N_B) - 6$ or $3(N_A + N_B) - 5$ degrees of vibrational freedom for a nonlinear or a linear complex, respectively. We note that

TABLE 6.1. Partition functions for different types of motion [Ref. 7,8]

Motion	Degrees of freedom	Partition function*	Order of magnitude
Translation	3	$\dfrac{(2\pi m k_B T)^{3/2}}{h^3}$ (per unit volume)	10^{25}–10^{26} cm^{-3}
Rotation (linear molecule)	2	$\dfrac{8\pi^2 I k_B T}{\sigma h^2}$	10–10^2
Rotation (nonlinear molecule)	3	$\dfrac{8\pi^2 \left(8\pi^3 I_A I_B I_C\right)^{1/2} (k_B T)^{3/2}}{\sigma h^3}$	10^2–10^3
Vibration (per normal mode)	1	$\dfrac{1}{1 - e^{-h\nu/k_B T}}$	1–10

Where   m = mass of molecule
         I = moment of inertia for linear molecule
I_A, I_B, and I_C = moments of inertia for a nonlinear molecule about three axes perpendicular to each other
         ν = normal-mode vibrational frequency
         k_B = Boltzmann constant
         h = Planck constant
         T = absolute temperature
         σ = symmetry number
*The power to which h appears is equal to the number of degrees of freedom.

one of these modes is different from the rest because it corresponds to the loose vibrational mode with little or no restoring force, thus product formation can occur upon its presence and this degree of vibrational freedom is associated with the reaction coordinate. Consequently, the factor from Table 6.1 corresponding to this loose vibration mode, where h is Planck's constant:

$$1 / \left(1 - e^{-h\nu/k_B T}\right) \tag{6.7}$$

is evaluated at the limit in which ν, the frequency of vibration, approaches zero, i.e.,

$$\lim_{\nu \to 0} \frac{1}{1 - e^{-h\nu/k_B T}} = \frac{1}{1 - (1 - h\nu/k_B T)} = k_B T / h\nu \tag{6.8}$$

if this exponential function is expanded and only the first term in the series is kept. This term is now factored out of $Q_{X\ddagger}$ to give

$$Q_{X\ddagger} = (k_B T / h\nu) Q^\ddagger \tag{6.9}$$

where $Q^\ddagger$ has $3(N_A + N_B) - 7$ or $3(N_A + N_B) - 6$ degrees of vibrational freedom for a nonlinear or a linear molecule, respectively. Substituting this into equation 6.1 and rearranging gives:

$$\nu C_{X\ddagger} = C_A C_B \frac{k_B T}{h} \frac{Q^\ddagger}{Q_A Q_B} e^{-E_0/RT} = r \tag{6.10}$$

where ν is the vibrational frequency in the activated complex along the reaction coordinate to form products. Because $C_{X\ddagger}$ is the concentration of the activated complex (the reactant), this determines the rate of reaction, r; consequently, the rate constant defined by

$$r \equiv k C_A C_B \tag{6.11}$$

is $\quad k = \dfrac{k_B T}{h} \dfrac{Q^\ddagger}{Q_A Q_B} e^{-E_0/RT} = A e^{-E_0/RT} \tag{6.12}$

from TST and is of the Arrhenius form, with A representing the pre-exponential factor. The term $\frac{k_B T}{h}$ is typically referred to as the universal frequency and has units of reciprocal time. Other derivations are possible and the same result is obtained [7].

Finally, it should be pointed out that this approach has assumed that an ideal system exists in which activity coefficients are unity. In a more general sense which also applies to nonideal systems, activities of species i, a_i, rather than concentrations, must be used; consequently, activity coefficients, γ_i, are introduced into equations 6.10 and 6.12 because $a_i = \gamma_i C_i$ [9].

A comparison of equations 6.1 and 6.12 shows that a modified equilibrium constant can be written for the formation of the activated complex, i.e., $K^\ddagger = \frac{k_B T}{h} K_\ddagger$, then

$$k = \frac{k_B T}{h} e^{-\Delta G_{\ddagger}^{o}/RT} = \frac{k_B T}{h} e^{\Delta S_{\ddagger}^{o}/R} e^{-\Delta H_{\ddagger}^{o}/RT} \tag{6.13}$$

Thus the rate constant k can also be expressed in terms of the changes in the standard free energy $\left(\Delta G_{\ddagger}^{o}\right)$, the standard enthalpy $\left(\Delta H_{\ddagger}^{o}\right)$ and the standard entropy $\left(\Delta S_{\ddagger}^{o}\right)$ when the activated complex is formed. There is a small difference if this is written in terms of the activation energy, E_a, associated with the rate constant in the Arrhenius expression represented by equation 6.12, rather than the enthalpy of formation, and equation 6.13 then becomes [7]:

$$k = \frac{(e^{1-\Delta n}) k_B T}{h} e^{-\Delta S_{\ddagger}^{o}/R} e^{-E_a/RT} \tag{6.14}$$

where Δn is the change in the number of moles when the activated complex is formed from the reactants ($\Delta n = -1$ for a bimolecular reaction, for example).

Even with this limited background, one can readily evaluate the role of a heterogeneous catalyst by comparing the rates of a homogeneous reaction and a catalyzed reaction between the same reactants. For example, let us choose the reaction between hydrogen and iodine, which was extensively studied by Bodenstein [10]. In the gas phase this reaction can be represented analogously to equations 6.1 and 6.2 as:

(6.3) $$H_2 + I_2 \underset{\ominus}{\overset{K^{\ddagger}}{\rightleftharpoons}} \left[\begin{array}{c} H \\ H \quad I \\ I \end{array} \right]^{\ddagger}$$

(6.4) $$\left[\begin{array}{c} H \\ H \quad I \\ I \end{array} \right]^{\ddagger} \xrightarrow{k} 2HI$$

The rate equation from transition state theory (TST) is:

$$r_{hom} = \frac{k_B T}{h} \frac{Q^{\ddagger}}{Q_{H_2} Q_{I_2}} e^{-E_{hom}/RT} C_{H_2} C_{I_2} = A_o e^{-E_{hom}/RT} C_{H_2} C_{I_2} \tag{6.15}$$

and it is emphasized again that the Q_i terms are per volume. Values calculated from the expressions in Table 6.1 give a value of $A_o = 2 \times 10^{14}\,cm^3\,mole^{-1}\,s^{-1}$ at 700 K, which is equal to the value obtained experimentally because k and E_{hom} were measured to be $64\,cm^3\,mole^{-1}\,s^{-1}$ and $40\,kcal\,mole^{-1}$, respectively [10]. Pre-exponential factors calculated from TST for homogeneous reactions are frequently in quite satisfactory agreement with experimentally determined values [7].

For a second-order heterogeneous reaction between H_2 and I_2 adsorbed on a surface, one can visualize the activated complex in chemical equation

6.3 being formed on a pair of neighboring active sites, S_p, and the reaction rate from TST per unit surface area is:

$$r_{het} = L_p \frac{k_B T}{h} \frac{Q_s^{\ddagger}}{Q_{H_2} Q_{I_2} Q_{S_p}} e^{-E_{het}/RT} C_{H_2} C_{I_2} = A_o' \; e^{-E_{het}/RT} C_{H_2} C_{I_2} \quad (6.16)$$

where $Q_s^{\ddagger}$ and Q_{S_p} are per unit surface area and L_p is the density of site pairs. If the reactants are strongly bound, then these latter two partition functions contain only high frequency vibrations and are thus near unity. If the ratio of these two rates is examined, one sees that

$$r_{het}/r_{hom} = \frac{L_p}{Q^{\ddagger}} e^{(E_{hom}-E_{het})/RT} = \frac{L_p}{Q^{\ddagger}} e^{\Delta E/RT} \quad (6.17)$$

Now, it has been mentioned before that the site density L (or L_p in this case) is typically near $10^{15} \, \text{cm}^{-2}$, and a reasonable value for $Q^{\ddagger}$ from Table 6.1 is $\sim 10^{27} \, \text{cm}^{-3}$, therefore:

$$r_{het}(\text{cm}^2)/r_{hom}(\text{cm}^3) \approx 10^{-12} e^{\Delta E/RT} \quad (6.18)$$

It is clear that a catalyst should lower the apparent activation energy if it is to enhance the reaction rate, even if the catalyst has a high specific surface area. If a catalyst has a relatively high surface area of $100 \, \text{m}^2 \, \text{g}^{-1}$ and its density is near unity, then there is about $10^6 \, \text{cm}^2$ per cm^3, and if $E_{hom} = E_{het}$:

$$r_{het}/r_{hom} \approx 10^{-6} \quad (6.19)$$

thus estimates of the decrease in activation energy required to enhance rates can easily range from 10–30 kcal mole^{-1}, depending on the actual specific surface area of the catalyst and the temperature range used, and such decreases, or greater, in activation energy are routinely observed [7].

6.2   The Steady-State Approximation (SSA)

It has been stated by Boudart that "the steady-state approximation (SSA) can be considered as the most important general technique of applied chemical kinetics" [9]. A formal proof of this hypothesis that is applicable to all reaction mechanisms is not available because the rate equations for complex systems are often impossible to solve analytically. However, the derivation for a simple reaction system of two first-order reactions in series demonstrates the principle very nicely and leads to the important general conclusion that, to a good approximation, the rate of change in the concentration of a reactive intermediate, X, is zero whenever such an intermediate is slowly formed and rapidly disappears.

For the reaction system just mentioned one has:

(6.5)
$$A \xrightarrow{k_1} X \xrightarrow{k_2} B$$

If C_A, C_X, and C_B are the concentrations of A, X and B at any time, t, and C_{A_o} is the concentration of A at t = 0 at which time $C_X = C_B = 0$, then mass balances on each of the three species give:

$$-\frac{dC_A}{dt} = k_1 C_A \tag{6.20}$$

$$\frac{dC_X}{dt} = k_1 C_A - k_2 C_X \tag{6.21}$$

and
$$\frac{dC_B}{dt} = k_2 C_X \tag{6.22}$$

with the obvious constraint that

$$C_{A_o} = C_A + C_X + C_B \tag{6.23}$$

Integration of these differential equations with the stated boundary conditions gives:

$$C_A/C_{A_o} = e^{-k_1 t}, \tag{6.24}$$

$$C_X/C_{A_o} = \left(\frac{k_1}{k_2 - k_1}\right)(e^{-k_1 t} - e^{-k_2 t}), \tag{6.25}$$

and

$$C_B/C_{A_o} = \frac{k_2(1 - e^{-k_1 t}) - k_1(1 - e^{-k_2 t})}{k_2 - k_1} \tag{6.26}$$

One can now consider two limiting cases: $k_1 \gg k_2$ and $k_1 \ll k_2$. In the former situation, the intermediate species X is rapidly formed but reacts slowly and equation 6.25 indicates that, after a short induction period of $e^{-k_1 t}$,

$$C_X = C_{A_o} e^{-k_2 t} \tag{6.27}$$

thus C_X approaches C_{A_o}, and C_B is formed according to a simple 1st-order rate expression. In contrast, in the latter situation after this induction period $e^{-k_2 t} \cong 0$ and, to a good approximation:

$$C_X = C_{A_o}(k_1/k_2)e^{-k_1 t} = C_A k_1/k_2 \tag{6.28}$$

thus $k_2 C_X = k_1 C_A$ and from equation 6.21 the SSA is obtained, i.e.,

$$dC_X/dt = d[X]/dt = 0 \tag{6.29}$$

This derivative for reactive intermediates is not to be integrated, because the result that C_X equals a constant is shown to be incorrect by equation 6.28. In addition, for these systems C_X must remain small compared to the concentrations of stable products and reactants. Finally, the induction period is typically very short compared to the overall reaction time; however, one can easily imagine that the induction period should consist of at least one turnover of the catalytic cycle not only to allow the concentrations of the reactive intermediates to be established and stabilized, but also to verify that a catalytic reaction is occurring [11]. Consequently, data should preferentially be acquired after this induction period, which is the inverse of the TOF and, because in the laboratory TOFs are frequently on the order of $10^{-2} - 10^{-3}$ s^{-1}, 2–15 minutes on stream might be required before the first sample is obtained that is representative of steady-state conditions.

The steady-state approximation has been applied for decades to closed systems such as homogeneous chain reactions, and an example of such an application can nicely demonstrate how valuable the SSA is. However, before this example is examined, a few comments about chain reactions are appropriate. Such a sequence of steps must consist of an initiation reaction to form the chain carriers, a chain propagation cycle consisting of at least two reactions in a closed sequence, and a termination reaction to remove chain carriers from the system. The rates of the steps in the propagation cycle are invariably much greater than those in the initiation and termination steps, thus product formation occurs primarily via this cycle. This is reflected in the term "chain length", which represents the average number of times the closed propagation cycle is repeated with the chain carrier once it is formed, thus it is the rate of the overall reaction divided by the rate of the initiation step, and this ratio can often have values far greater than 10^6 [12]. Now consider the gas-phase ozone decomposition reaction that is described in Illustration 6.1.

Illustration 6.1 – Homogeneous Ozone Decomposition – Application of the SSA

The gas-phase ozone decomposition reaction, $2O_3 \longrightarrow 3O_2$, can be catalyzed by chlorine at low temperature [13]. The experimental rate equation was reported to be:

$$r = \frac{d\xi}{Vdt} = \frac{1}{\nu_i}\frac{dC_i}{dt} = \frac{1}{3}\frac{C_{O_2}}{dt} = -\frac{1}{2}\frac{dC_{O_3}}{dt} = kC_{Cl_2}^{1/2}C_{O_3}^{3/2} = k[Cl_2]^{1/2}[O_3]^{3/2} \quad (1)$$

and the following chain reaction sequence has been proposed to describe this reaction:

(1) $Cl_2 + O_3 \xrightarrow{k_1} ClO \cdot + ClO_2 \cdot$ (Initiation)

(2) $ClO_2 \cdot + O_3 \xrightarrow{k_2} ClO_3 \cdot + O_2$ ⎫

(3) $ClO_3 \cdot + O_3 \xrightarrow{k_3} ClO_2 \cdot + 2O_2$ ⎬   (Propagation)

(4) $2 \text{ ClO}_2 \cdot \xrightarrow{k_4} \text{Cl}_2 + 2 \text{ O}_2$ $\Bigg\}$ (Termination)

(5) $2 \text{ ClO} \cdot \xrightarrow{k_5} \text{Cl}_2 + \text{O}_2$

The derivation of the rate expression based on this model can be approached in two equivalent ways, but to initiate either one, the rate must first be defined. In this case it is either that of ozone disappearance or that of dioxygen formation, noting from equation 1 that

$$\frac{d[O_2]}{dt} = -\frac{3}{2}\frac{d[O_3]}{dt} \tag{2}$$

If the former is selected, then from steps 2 and 3 in the reaction sequence, i.e., the chain propagation cycle:

$$r = d[O_2]/dt \cong r_2 + 2r_3 = k_2[ClO_2 \cdot][O_3] + 2k_3[ClO_3 \cdot][O_3] \tag{3}$$

The contribution to O_2 formation from the two termination steps is insignificant because chain lengths are large ($r_2, r_3 \gg r_1, r_4, r_5$), thus the latter two rates are negligible and can be ignored. The rate for each of these elementary steps can be written just as it appears based on the law of mass action. In one approach to eliminate the unknown free radical concentrations from these rate expressions, the SSA is now applied to each of the reactive intermediates within the propagation sequence. Thus, for $ClO_3 \cdot$ which is formed in step 2 and consumed in step 3:

$$\frac{d[ClO_3 \cdot]}{dt} = k_2[ClO_2 \cdot][O_3] - k_3[ClO_3 \cdot][O_3] = 0 \tag{4}$$

and we learn that $r_2 = r_3$. The SSA for $ClO_2 \cdot$ gives the same result. Therefore, equation (3) can be simplified to:

$$r = 3r_2 = 3k_2[ClO_2 \cdot][O_3] \tag{5}$$

To remove $[ClO_2 \cdot]$ from this expression and obtain a rate law containing only the concentrations of known, stable compounds such as products and reactants, the SSA is applied again based on the assumption that at steady-state, the rate of initiation of chain carriers, in this case $ClO_2 \cdot$, must equal the rate of their termination, i.e.,

$$r_i = r_t = k_1[Cl_2][O_3] = 2k_4[ClO_2 \cdot]^2 \tag{6}$$

Note that only the free radicals involved in the propagation sequence are important, thus step 5 is irrelevant. Also note the stoichiometric coefficient of 2 required in step 4 because:

$$r_t = \frac{1}{V}\frac{d\xi}{dt} = \frac{1}{\nu_i}\frac{d[i]}{dt} = \frac{-1}{2}\frac{d[ClO_2 \cdot]}{dt} = k_4[ClO_2 \cdot]^2 \tag{7}$$

thus

$$\frac{d[ClO_2 \cdot]}{dt} = 2k_4[ClO_2 \cdot]^2 \tag{8}$$

From equation 6 one obtains:

$$[ClO_2 \cdot] = (k_1/2k_4)^{1/2}[Cl_2]^{1/2}[O_3]^{1/2} \tag{9}$$

and substitution of this expression into equation 5 gives the final rate equation, which is consistent with the experimental results:

$$r = 3k_2(k_1/2k_4)^{1/2}[Cl_2]^{1/2}[O_3]^{3/2} \tag{10}$$

Thus the apparent rate constant in equation 1 is composed of three elementary-step rate constants, i.e., k_1, k_2 and k_4.

An alternate equivalent approach to eliminate the unknown reactive intermediate concentrations is to apply the SSA to each intermediate within the entire reaction sequence, i.e.,

$$\frac{d[ClO_3 \cdot]}{dt} = k_2[ClO_2 \cdot][O_3] - k_3[ClO_3 \cdot][O_3] = 0 \tag{11}$$

which is identical to equation 4, and

$$\frac{d[ClO_2 \cdot]}{dt} = k_1[Cl_2][O_3] - k_2[ClO_2 \cdot][O_3] + k_3[ClO_3 \cdot][O_3] - 2k_4[ClO_2 \cdot]^2 = 0 \tag{12}$$

Because equation 11 and equation 12 each sums to zero, they can be added or subtracted with impunity, and adding the two gives:

$$k_1[Cl_2][O_3] - 2k_4[ClO_2 \cdot]^2 = 0 \tag{13}$$

which is the same as equation 6. Although usually not necessary, the contributions to the overall rate by the initiation and/or termination steps can be included for the greatest accuracy. In similar fashion, for heterogeneous catalytic reactions which do not have a RDS, the SSA will be the tool of choice to eliminate the concentrations of unknown reactive intermediates.

6.3   Heats of Adsorption and Activation Barriers on Metal Surfaces: BOC–MP/UBI-QEP Method

The capability to calculate heats of chemisorption from some model would be an enormous benefit, not only in experimental adsorption studies, which could provide results to check the model, but also in heterogeneous catalytic studies involving surface intermediates. This has long been recognized, and early efforts were made to accomplish this with metals based on various measurable properties, but they met with limited success [14,15]. Since the advent of modern surface science techniques beginning about forty years ago, a significant amount of data has been gathered related to the bond strength of various atoms adsorbed on different metals. Using values for atomic adsorption acquired from molecular adsorption on different transition metal crystal planes, Shustorovich initially constructed a mathematical formalism based on thermodynamics, namely, the bond-order conservation-Morse potential (BOC-MP) model [16–18]. Utilizing both reported and estimated heats of atomic chemisorption on transition metal surfaces, this model, which contains both theoretical and empirical aspects, allows the calculation of not only heats of adsorption of polyatomic species, but also activation barriers for their decomposition and recombination on surfaces. Consequently, for many reactions on metal surfaces, various reaction pathways can be proposed and the energetics associated with each can be estimated.

Very recently, Shustorovich and coworkers have extended the conceptual framework of this initial BOC-MP model to prove that the quadratic exponential potential (QEP), expressed in terms of a unity bond index (UBI) after normalization, provides a general, accurate description of any two-center, quasi-spherical interaction. As a result, the formalism has been renamed the UBI-QEP method [19–22]. This recent work includes a newly developed formalism to calculate the heat of adsorption of polyatomic molecules, such as ethylene and acetylene, without bond energy partitioning [19,22]. The reader is invited to study these latter papers to learn the most recent and accurate applications of this approach, which will be referred to as the BOC-MP/UBI-QEP method in this chapter. In this section, only an overview of the initial BOC-MP method will be provided, and greater detail is provided in these publications of Shustorovich [16–18].

6.3.1 Basic BOC-MP/UBI-QEP Assumptions

It is convenient to first establish some definitions, and these are listed in Table 6.2. As discussed in Chapter 5, as an adsorbate approaches a surface, changes occur in the potential energy of the system and one must relate this energy, E_{pot}, to the distance, r, from the surface. It is certainly anticipated that the migration and dissociation of an adsorbate, X, on the surface involve changes in the coordination mode, $M_n - X$, and the $M-X$ distance, r, where M is a metal atom and n is the coordination number. Assuming quasi-spherical interactions, a two-center M-A bond index, x, in the form of Pauling's bond order [23], is defined as

$$x = e^{-(r-r_0)/a} \tag{6.30}$$

where r_0 and a are constants, with the former representing the equilibrium distance when the bond index (order) is unity. For two-center interactions, the simplest relationship for potential energy containing both attractive and repulsive forces is a quadratic exponential potential (QEP) in the form of the Morse potential [24]:

TABLE 6.2. Definitions associated with the BOC-MP/UBI-QEP Model [Ref. 16–18]

X	–	Adsorbate, general
A	–	Atomic adsorbate
AB	–	Molecular adsorbate, diatomic
M	–	Metal
Q_A	–	Experimental heat of adsorption for atom, A
Q_{oA}	–	Maximum 2-center (M-A) bond energy
D_{AB}	–	Gas-phase dissociation energy of molecule A-B
Q_{AB}	–	Molecular heat of adsorption of AB
n	–	Coordination number of atom on surface
n'	–	Coordination number of adsorbed molecule or radical

$$-Q(x) = E_{pot}(x) = -Q_o(2x - x^2) \qquad (6.31)$$

where Q_o is the M-A equilibrium bond energy. The total potential energy, $E_{pot}(x)$, has only one minimum, which occurs when the bond order $x = 1$ and $r = r_o$, as depicted in Figure 6.2.

To move from two-center (M−A) to many-center (M_n− A) interactions, where n is the number of metal atoms coordinated with A, it is assumed that the sum of all two-center M-A interactions constitutes the total interaction for the M_n− A system, i.e.,

$$Q(n) = \sum_{i=1}^{n} Q_i \qquad (6.32)$$

and

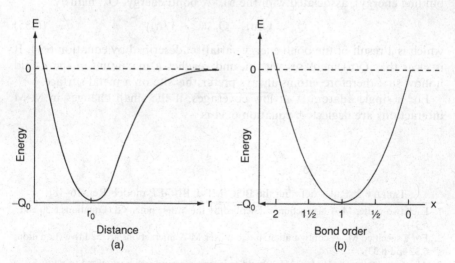

FIGURE 6.2. Presentations of the Morse potential: (a) E versus r and (b) E versus x. See Eq. 6.30 and Eq. 6.31 in the text. (Reprinted from ref. 17, copyright © 1986, with permission from Elsevier)

$$x(n) = \sum_{i=1}^{n} x_i \qquad (6.33)$$

It is next assumed that the total bond index (order), x, does not change with n, and x is conserved, hence bond-index (order) conservation. Because, by definition, the equilibrium two-center bond index is $x_o = 1$ for $n = 1$, then x is also normalized to unity for any $n \geq 1$, i.e.,

$$x(n) = x = \sum_{i=1}^{n} x_i = x_o = 1 \qquad (6.34)$$

The final assumption is that n is limited to nearest-neighbor metal atoms. For instance, for A adsorbed on an fcc (100) surface the maximum value of n is 4, which occurs for adsorption in the 4-fold hollow site, while on an fcc (111) surface n is a maximum of 3 in the 3-fold hollow site, and on either surface $n = 2$ in a bridge site and $n = 1$ in an on-top site. Table 6.3 summarizes the assumptions that govern all calculations using the BOC-MP model, which involve only algebraic relationships.

6.3.2 Heats of Atomic Chemisorption

In the subsequent BOC-MP/UBI-QEP relationships, the basic energetic parameter is the Morse constant Q_o in equation 6.31, which corresponds to the maximum two-center M-A bond energy, Q_{oA}, for atom A adsorbed on an on-top site. This value is not directly obtainable, but it can be readily determined from the experimental heat of adsorption value, Q_A (the atomic binding energy), associated with the M_n-A bond energy, Q_n, namely

$$Q_A = Q(n) = Q_{oA}(2 - 1/n) \qquad (6.35)$$

which is a result of the bond energy variation described by equation 6.34. It is clear that Q(n) increases with n, and reaches a maximum in the n-fold hollow site; therefore, atoms always prefer this site on a metal surface.

For a single adatom A at low coverages, if the small changes in M-M interactions are neglected, equation 6.34 is

TABLE 6.3. Rules defining the BOC-MP/UBI-QEP model [Ref. 16–18]

1. Each two-center M-A interaction is described by the Morse potential (Equations 6.30 and 6.31).
2. For a specified M_n-A configuration, n two-center M-A interactions are additive (Equations 6.32 and 6.33).
3. Along any pathway the total M_n-X bond order is conserved and normalized to unity (Equations 6.34, 6.39, 6.40).
4. For a specified M_n-A configuration, n is constrained to nearest neighbor atoms.

$$\sum_{i=1}^{n} x_{A_i} = 1 \qquad (6.36)$$

and equations 6.31 and 6.32 give

$$Q(n) = Q_{oA} \sum_{i=1}^{n} \left(2x_{A_i} - x_{A_i}^2 \right) = Q_{oA} \left(2 - \sum_{i=1}^{n} x_{A_i}^2 \right) \qquad (6.37)$$

The maximum value for $Q(n)$ occurs for equivalent M-A interactions with an equal bond order of $1/n$. Then equation 6.30 gives $r_n = r_o - a \ln x$, which results in equation 6.35. Consequently, the desired value is obtained:

$$Q_{oA} = Q_A / (2 - 1/n) \qquad (6.38)$$

where n is the coordination number of an atom on a metal surface plane. Table 6.4 illustrates that $Q(n)$ changes significantly as the atom moves from an on-top site to a hollow site, but does not vary markedly after n reaches 3 or more. Experimental heats of chemisorption for H, O, N and C atoms are listed in Table 6.5.

6.3.3   Heats of Molecular Chemisorption

For chemisorption of a molecule AB, from equation 6.34 bond-order conservation gives

$$\sum_{i=1}^{n} (x_{A_i} + x_{B_i}) + x_{AB} = 1 \qquad (6.39)$$

which relates the heat of adsorption, Q_{AB}, to the heats of adsorption of the atoms, Q_A and Q_B, and to the gas-phase dissociation energy, D_{AB}, of the molecule. However, this equation still has too many variables to allow an analytical solution, so it must be simplified. One way is to neglect certain metal-adsorbate interactions which have minor contributions, and this is

TABLE 6.4. Q_o versus Q_n for atomic chemisorption M_n-A (from ref. 16)

n	Site	Surface	Q_n/Q_o [a]
1	On-top		1.00
2	Bridge C_{2v}		1.50
3	Hollow C_{3v}	hcp(001)	1.67
		fcc(111)	
		bcc(110)	
4	Hollow C_{4v}	fcc(100)	1.75
5	Hollow C_{4v}	bcc(100)	1.80
6–9 [b]		Stepped, kink	1.83–1.89
12			1.92

[a] $Q_n/Q_o = 2 - 1/n$ (equation 6.35 or 6.38)
[b] Possible high coordinations on rough surfaces.

TABLE 6.5. Experimental heats of atomic chemisorption, Q_A, on some metal surfaces[a]

Metal surface	Atom			
	H	O	N	C
W(110)	68	~125	155	(200)[b]
Fe(110)	64	(118)[c]	140	(200)[b]
Ru(001)	67	100	—	—
Rh(111)	61	102	—	—
Ir(111)	58	93	127	—
Ni(111)	63	115	135	171
Pd(111)	62	87	130	(160)[b]
Pt(111)	61	85	116	(150)[b]
Cu(111)	56	103[d]	—	(120)[b]
Ag(111)	52	80	—	110[e]
Au(110)	46[e]	75[e]	—	108[e]

[a] In kcal mole^{-1} (from ref. 16).
[b] Assumed values (see ref. 16).
[c] For polycrystalline Fe.
[d] For a polycrystalline surface.
[e] See refs. 19, 20.

determined by the geometry of the $M_{n'} - AB$ configuration. Another route is to use the effective group terms, such as the effective atomic bond order $x_A = \sum_{i=1}^{n} x_{A_i}$. For the latter case, equation 6.39 is rewritten as:

$$x_A + x_B + x_{AB} = 1 \qquad (6.40)$$

The difference between the last two equations is that different Morse constants are used to describe the interaction of an atom A in a molecule AB with a metal surface, i.e., the two-center M-A energy Q_{oA} for equation 6.39 and the polycenter $M_{n'} - AB$ energy Q_A for equation 6.40. Consequently, a choice must be made about which molecules are best described by each representation, and this essentially depends on whether chemisorption is "weak" or "strong".

"Weak" chemisorption would be chosen when the A-B bond is strong (large D_{AB}) and all valences are satisfied, thus equation 6.39 would be used for molecules such as CO, H_2, O_2, NH_3 and H_2O. The bonding of these molecules is relatively insensitive to the coordination site, thus they prefer on-top or, possibly, bridging sites. If AB is monocoordinated (η^1) via one atom to a surface (an on-top site) with the A end down, the M-B interactions can be neglected, and this on-top M-A-B coordination is described by the term $\eta^1 \mu_1$, where the subscript on μ represents n′, the number of metal atoms involved. For the η^1 configuration, the molecule is coordinated to n′ M atoms via the atom with the highest atomic heat of adsorption, i.e., for M-A-B, $Q_A > Q_B$. Then, for $\eta^1 \mu_{n'}$ bonding

$$Q_{AB} = Q_{oA}^2 / \left(\frac{Q_{oA}}{n'} + D_{AB} \right) \quad \left(\text{for } D_{AB} > \frac{n'-1}{n'} Q_{oA} \right) \qquad (6.41)$$

If a molecule adsorbs in a dicoordinated mode (η^2) via two atoms, i.e., a bridge site (μ_2), then for this $(\eta^2\mu_2)$ configuration, where A and B can be either atoms or groups treated as quasi-atoms:

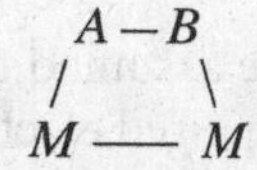

$$
\begin{array}{c}
A - B \\
/ \quad\quad \backslash \\
M \longrightarrow M
\end{array}
$$

the equation for the bonding energy on the surface is

$$Q_{AB} = \frac{ab(a+b) + D_{AB}(a-b)^2}{ab + D_{AB}(a+b)} \tag{6.42}$$

where

$$a = Q_{oA}^2(Q_{oA} + 2Q_{oB})/(Q_{oA} + Q_{oB})^2 \tag{6.43}$$

and

$$b = Q_{oB}^2(Q_{oB} + 2Q_{oA})/(Q_{oA} + Q_{oB})^2 \tag{6.44}$$

For a bridge-bonded homonuclear molecule A_2, equations 6.43 and 6.44 reduce to:

$$a = b = a_o = (3/4)Q_{oA} \tag{6.45}$$

and equation 6.42 simplifies to

$$Q_{A_2} = (9/2)Q_{oA}^2/(3Q_{oA} + 8D_{A_2}) \tag{6.46}$$

If a multiatom molecule is dicoordinated at a bridge site in a chelated form, i.e.,

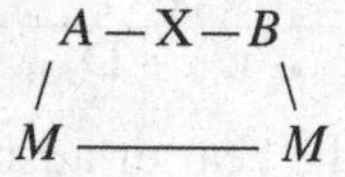

$$
\begin{array}{c}
A - X - B \\
/ \quad\quad\quad \backslash \\
M \longrightarrow M
\end{array}
$$

where atoms A and B are linked by atom X, and the modified Morse constants are now

$$Q'_{oA(X)} = a' = Q_{oA}^2/(Q_{oA} + D_{AX}) \tag{6.47}$$

and

$$Q'_{oB(X)} = b' = Q_{oB}^2/(Q_{oB} + D_{BX}) \tag{6.48}$$

which, when substituted into equations 6.43 and 6.44 give

$$a_X = a'^2(a' + 2b')/(a' + b')^2 \tag{6.49}$$

and

$$b_X = b'^2(b' + 2a')/(a' + b')^2 \tag{6.50}$$

Thus, the heat of chemisorption of this chelated species is

$$Q_{AB(X)} = a_X + b_X \qquad (6.51)$$

or

$$Q_{A_2(X)} = 2a_X \qquad (6.52)$$

"Strong" chemisorption would be assumed to occur for surface species such as molecular radicals in which unpaired electrons retain most of their atomic character, and the adsorption pattern would resemble that for atoms, which includes a distinct preference for n-fold hollow sites. Examples would include radicals like CH, CH_2, NH, OH and OCH_3. In this case for mono-coordination $(\eta^1 \mu_{n'})$, such as $M_{n'} - AB$, the Morse constants are better represented by the experimental heats of atomic chemisorption, Q_A and Q_B, and the use of equation 6.40 provides the following respective analogues for equations 6.41 and 6.52:

$$Q_{AB} = Q_A^2/(Q_A + D_{AB}) \qquad (6.53)$$

and

$$Q_{A_2(X)} = 2a_X = (3/2)a' = (3/2)Q_A^2/(Q_A + D_{AX}) \qquad (6.54)$$

Other analogies can be obtained in similar fashion [16].

Finally, intermediate bond strengths might be anticipated for monovalent AB radicals in which A is a tri- or tetra-valent atom, such as the N or C atom in NH_2, CH_3 or HCO [16]. In this situation, the heat of adsorption can be obtained by averaging the two limiting calculations, i.e., equations 6.41 and 6.53. Thus for a $(\eta^1 \mu_{n'})M_{n'} - AB$ configuration

$$Q_{AB} = 1/2\left[\frac{Q_{oA}^2}{(Q_{oA}/n') + D_{AB}} + \frac{Q_A^2}{Q_A + D_{AB}}\right] \qquad (6.55)$$

Some calculated values for the heat of adsorption of diatomic molecules, Q_{AB}, are listed in Table 6.6 and compared to experimental values obtained from the literature [16]. The effect of the type of coordination and the bonding end of a molecule can readily be observed. Illustration 6.2 provides some sample calculations resulting in these values for "weak" chemisorption. Additional examples are given elsewhere [16–18].

Illustration 6.2 – BOC-MP/UBI-QEP Calculations of Molecular Heats of Chemisorption on Metal Surfaces

By the definition used in this section, adsorption of molecules would be considered "weak" chemisorption; therefore, the appropriate equation to use is equation 6.41 along with equation 6.38. For example, assume a $\eta^1 \mu_1$ configuration for CO adsorbed carbon end down on an on-top site (M-CO), which is expected because $Q_c > Q_o$ (See Table 6.5).

a) On a close-packed surface, n = 3, thus if CO adsorption on Ni (111) is chosen:

TABLE 6.6. Heats of molecular chemisorption Q_{AB}: diatomic molecules[a]

| System | Coord. type | Experimental values of [a] | | | Q_{AB} Calculated | | Exper. |
		Q_A	Q_B	D_{AB}	M-AB	(M-BA)	M-AB
CO/Ni(111)	$\eta^1\mu_1$	171	(115)	257	29^b	$(15)^b$	27
CO/Fe(110)	$\eta^1\mu_1$	200	(125)	257	38	(17)	36
CO/M	$\eta^1\mu_1$	150–200	(85–125)	257	25–38^b	$(8$–$17)^b$	26–40
NO/Pt(111)	$\eta^1\mu_2$	116	(85)	151	26^b	$(15)^b$	27
NO/Pd(111)	$\eta^1\mu_2$	130	(87)	151	32^b	$(15)^b$	31
O$_2$/Pt(111)	$\eta^2\mu_2$	85	85	119	11^c	—	9
O$_2$/Ag(110)	$\eta^2\mu_2$	80	80	119	10^c	—	10
N$_2$/Pt(111)	$\eta^2\mu_2^d$	116	116	226	11^c	—	9
N$_2$/Ir(110)	$\eta^2\mu_2^d$	127	127	226	13^c	—	11
N$_2$/Ni(100)	$\eta^2\mu_2^d$	135	135	226	14^c	—	11
N$_2$/Ni(110)	$\eta^2\mu_2^d$	135	135	226	14^c	—	11
N$_2$/Fe(111)	$\eta^2\mu_2^e$	140	140	226	15^e	—	8

[a] See ref. 16 for details. All energies in kcal mole^{-1}.
[b] From equation 6.41.
[c] From equation 6.46.
[d] Assuming the same energy as in the experimentally observed coordination $\eta^1\mu_{n'}$.
[e] Coexists with the coordination $\eta^1\mu_{n'}$ (ref. 16).

$$Q_{oC} = Q_C/(2 - 1/n) = \frac{Q_C}{(2 - 1/3)} = 3/5 Q_C \tag{1}$$

and, from Tables 6.5 and 6.6

$$Q_{CO} = \frac{Q_{oC}^2}{\frac{Q_{oC}}{n'} + D_{CO}} = \frac{(0.6)^2(171)^2}{(0.6)(171) + 257} = 29 \,\text{kcal mole}^{-1} \tag{2}$$

b) For $\eta^1\mu_1$ adsorption on Cu(111):

$$Q_{CO} = \frac{(0.6)^2(120)^2}{(0.6)(120) + 257} = 16 \,\text{kcal mole}^{-1} \tag{3}$$

c) If a surface with a 4-fold hollow site is selected, n = 4, so for $\eta^1\mu_1$ CO adsorption on Ni (100):

$$Q_{oC} = O_C/(2 - 1/4) = 4/7 Q_C \tag{4}$$

and

$$Q_{CO} = \frac{(4/7)^2(171)^2}{(4/7)(171) + 257} = 27 \,\text{kcal mole}^{-1} \tag{5}$$

d) If a bridge-bonding configuration $(\eta^1\mu_2)$ is desired (n' = 2); for example CO on Ni (111), then equation 6.41 is:

$$Q_{CO} = \frac{(0.6)^2(171)^2}{\frac{(0.6)(171)}{2} + 257} = 34 \,\text{kcal mole}^{-1} \tag{6}$$

e) If a $\eta^1\mu_1$ adsorption mode is assumed for CO, but with the O end interacting with the surface; for example on Ni(111), the heat of chemisorption becomes:

$$Q_{CO} = \frac{(0.6)^2(115)^2}{(0.6)(115) + 257} = 15\,\text{kcal mole}^{-1} \tag{7}$$

This comparison clearly shows why CO chemisorbs C end down on metal surfaces.

f) A different adsorption configuration can be examined; for example, nondissociative O_2 adsorption on a Pt(111) surface ($\eta^2\mu_2$), and for this homonuclear molecule equation 6.46 would be applicable:

$$Q_{O_2} = \frac{(9/2)Q_{oO}^2}{3Q_{oO} + 8D_{O_2}} = \frac{(9/2)(0.6)^2(85)^2}{3(0.6)(85) + 8(119)} = 11\,\text{kcal mole}^{-1} \tag{8}$$

g) If a similar $\eta^2\mu_2$ configuration is assumed for CO to represent tilted or

parallel adsorption, $\begin{array}{c}\text{CO}\\ \diagup \quad \diagdown \\ \text{M} \text{---} \text{M}\end{array}$, then equation 6.42 would be utilized. A

comparison between equations 6.41 and 6.42 for the late transition metals shows that η^1 is favored over η^2, but the energy difference is not large ($\Delta Q < 5\,\text{kcal mole}^{-1}$) [16].

One can now examine the chemisorption of polyatomic molecules. The first consideration is the orientation of the molecule on the surface, which will be determined by the molecular orbitals and the degree of saturation of the bonds. For example, NH_3 would adsorb N end down, most likely in a $\eta^1\mu_1$ mode because of the large D_{NH_3} value for the N-H triple bond. With CH_3OH, adsorption via the O end of the molecule is expected. Illustration 6.3 shows the calculations associated with determining the heat of adsorption for methanol. The second consideration is the partitioning of bond energy within the molecular structure because once the M-AB two-center interaction is established (where B can represent one or more atoms or groups), the total energy of all bonds formed by atom A, D_{AB}, is required. In the BOC-MP/UBI-QEP model, D_{AB} is defined as the difference between the total gas-phase energy in the AB molecule and the dissociated fragments A and B. The only constraint with this approach is that such fragmentation must be endothermic, a situation which almost always prevails [16].

Illustration 6.3 – BOC-MP Calculation of the Heat of Chemisorption for CH_3OH on Pt(111)

Methanol is expected to adsorb O end down on a metal surface in a $\eta^1\mu_1$ configuration because all bonds are saturated. The bond energy D_{AB} associated with only the atom coordinated to the metal surface must now be determined. This can be accomplished by first using the following thermo-dynamic cycle to calculate the total bond energy in the molecule:

$$\text{(1)} \qquad \begin{array}{c} \text{C}\cdot \ + \ 4\text{H}\cdot \ + \ \text{O}\cdot \ \underline{\Delta\text{H}^\circ} \ \ \text{CH}_3\,\text{OH}_{(g)} \\[4pt] \Delta\text{H}^\circ_{f_\text{C}\cdot} \Big\uparrow \ \Delta\text{H}^\circ_{f_\text{H}\cdot} \Big\uparrow \ \Delta\text{H}^\circ_{f_\text{O}\cdot} \Big\uparrow \qquad \diagup\Delta\text{H}^\circ_f \\[4pt] \text{C}_{(s)} + 2\text{H}_{2\,(g)} + 1/2\text{O}_{2\,(g)} \end{array}$$

Because enthalpy is a state property and depends only on the initial and final states, two routes can be proposed to form CH_3OH, as shown in reaction 1, i.e.,

$$\Sigma\Delta\text{H}^\circ_{f_i}(\text{atoms}) + \Delta\text{H}^\circ = \Delta\text{H}^\circ_f, \qquad (1)$$

where $\Delta\text{H}^\circ_{f_i}$ is the standard enthalpy of formation of species i, thus the total bond energy in the molecule is $D_{Total} = \Delta\text{H}^\circ$, and

$$D_{Total} = \Sigma\Delta\text{H}^\circ_{f_i}(\text{atoms}) - \Delta\text{H}^\circ_f = D_{CH_3OH} \qquad (2)$$

One can obtain values for the heat of formation of gas-phase atoms and the standard enthalpy of formation of molecular species from various sources such as the CRC Handbook [25]. From this source one obtains enthalpies (in kcal mole^{-1}) for the various steps:

$$\Delta\text{H}^\circ_f(\text{C} \longrightarrow \text{C}\cdot) = 171.3$$
$$2\Delta\text{H}^\circ_f(\text{H}_2 \longrightarrow 2\,\text{H}\cdot) = 2[(2)(52.1)]$$
$$0.5\Delta\text{H}^\circ_f(\text{O}_2 \longrightarrow 2\text{O}\cdot) = 0.5(119.2)$$
$$-\Delta\text{H}_{f(CH_3OH)^\circ} = \frac{-(-48.0)}{487.3\,\text{kcal mole}^{-1}} = D_{CH_3OH} \qquad (3)$$

One must determine the bond energy associated with just the two O–X bonds, because $D_{HOCH_3} = D_{AB}$, and this can be determined if the total bond energy associated with the three C–H bonds in the CH_3 group, D_{CH_3}, is known. This latter value is determined from a thermodynamic cycle to form $CH_3\cdot$ analogous to the one above, i.e., from ref. [25]: $D_{CH_3\cdot} = \Delta\text{H}^\circ_{f_\text{C}} + 3\Delta\text{H}^\circ_{f_\text{H}} - \Delta\text{H}^\circ_{f_{(CH_3\cdot)}}$, i.e.,

$$\Delta\text{H}^\circ_{f_\text{C}\cdot} = 171.3$$
$$3\Delta\text{H}^\circ_{f_\text{H}\cdot} = 3(52.1)$$
$$\Delta\text{H}^\circ_{f(CH_3\cdot)} = \frac{-(+34.8)}{292.8\,\text{kcal mole}^{-1}} = D_{CH_3\cdot} \qquad (4)$$

Consequently, the energy associated only with the bonds involving the O atom is the difference between the total bond energy in the molecule and that in the three C-H bonds, i.e., D_{OH} plus D_{OC} is

$$D_{H-O-CH_3} = D_{Total} - D_{CH_3\cdot} = 487 - 293 = 194\,\text{kcal mole}^{-1} \qquad (5)$$

Then, for $\eta^1\mu_1$ CH_3OH adsorption on Pt(111) with the O end down, equation 6.41 ($n' = 1$) gives, after using equation 6.38 for oxygen:

$$Q_{CH_3OH} = Q_{oO}^2/(Q_{oO} + D_{H-O-CH_3}) = \frac{(0.6)^2(85)^2}{(0.6)(85) + 194} = 11 \, kcal \, mole^{-1} \quad (6)$$

Using the thermodynamic cycle employed in Illustration 6.3, total bond energies for a number of gas-phase molecules and radicals can be calculated, and some of these are listed in Table 6.7 along with their heats of chemisorption on the close-packed Ni, Pd, Pt and Fe/W surfaces. With the values contained in Table 6.7, bond strengths for the "strong" chemisorption associated with surface radical groups can readily be calculated using equation 6.53. For example, from Table 6.7a the binding energy of a CH group on Pt(111) would be

$$Q_{CH} = \frac{Q_C^2}{(Q_C + D_{CH})} = \frac{(150)^2}{(150 + 81)} = 97 \, kcal \, mole^{-1} \quad (7)$$

and for an OH group on Pd(111), the heat of chemisorption would be

$$Q_{OH} = \frac{(87)^2}{87 + 102} = 40 \, kcal \, mole^{-1} \quad (8)$$

TABLE 6.7a. Heats of chemisorption (Q) and total bond energies in the gas phase (D) and chemisorbed (D + Q) states on the platinum-group metals[a]

Species	D[b]	Ni(111)		Pd(111)		Pt(111)	
		Q	D + Q	Q	D + Q	Q	D + Q
C	—	171	171	160	160	150	150
CH	81	116	197	106	187	97	178
CH$_2$	183	83	266	75	258	68	251
CH$_3$	293	48	341	42	335	38	331
CH$_4$	398	6	404	6	404	6	404
H	—	63	63	62	62	61	61
O	—	115	115	87	87	85	85
OH	102	61	163	40	142	39	141
OH$_2$	220	17	237	10	230	10	230
OCH$_3$	383	65	448	43	426	41	424
CH$_3$OH	487	18	505	11	498	11	498
CO	257	27	284	34	291	32	289
HCO	274	50	324	44	318	40	314
H$_2$CO	361	19	380	12	373	11	372

[a] See ref. 16 for details. All energies in kcal mole^{-1}.
[b] Ref. 16 and ref. 25.

Illustration 6.4 – BOC-MP Calculation of the Heat of Chemisorption for Symmetric Polyatomic Molecules

Finally, the heat of chemisorption of symmetric polyatomic molecules should be briefly mentioned because equation 6.46 can be used to calculate this value, but one must remember that A now represents a molecular group, not an atom, and D_{A_2} represents the total bond energy contained in that molecular group and thus involves bond-energy partitioning. For example, let us use the earlier BOC-MP approach and consider ethylene adsorption on a metal, which would have a $\eta^2\mu_2$ configuration. Now A represents the -CH_2 group, and $D_{(CH_2)_2} = D_{(A)_2}$ is the energy associated with the -CH_2 half of the molecule. To get the energy in the C = C bond, $D_{C=C}$, from Table 6.7

$$D_{C=C} = D_{C_2H_4} - 2D_{CH_2} = 538 - 2(183) = 172\,\text{kcal mole}^{-1} \qquad (1)$$

TABLE 6.7b. Total bond energies in the gas-phase (D) and chemisorbed (D + Q) states on some transition metals[a]

C_2H_x	$D^b_{C_2H_x}$	$Q_{C_2H_x}{}^c$			$D_{C_2H_x} + Q_{C_2H_x} + (8-x)Q_H{}^d$		
		Fe/W	Ni	Pt	Fe/W	Ni	Pt
$H_3C - CH_3$	674	6	5	5	812	805	801
$H_3C - CH_2$	576	64	49	39	838	814	798
$H_3C - CH$	466[e]	107	85	70	837	803	780
$H_2C = CH_2$	538	20	15	12	822	805	794
$H_3C - C$	376[f]	141	115	97	847	806	778
$H_2C = CH$	421[g]	71	55	44	822	791	770
$H_2C = C$	348[h]	110	87	71	854	813	785
$HC \equiv CH$	392	25	18	14	813	788	772
$HC \equiv C$	259[i]	106	84	69	827	784	755
$CH_3 + CH_3$	586	124	96	76	842	808	784
$CH_3 + CH_2$	476	166	131	106	840	796	765
$CH_3 + CH$	374	204	164	135	842	790	753
$CH_3 + C$	293	262	219	188	885	827	786
$CH_2 + CH_2$	366	208	166	136	838	784	746
$CH_2 + CH$	264	246	199	165	840	778	734
$CH_2 + C$	183	304	254	218	883	815	767
$CH + CH$	162	284	232	194	842	772	722
$CH + C$	81	342	287	247	885	809	755
$C + C$	0	400	342	300	928	846	788
$CH_4 + CH_4$	796	14	12	12	810	808	808

[a] All energies are in kcal mole^{-1}. The parameters used: Q_C = 150, 171, and 200 and Q_H = 61, 63, and 66 for Pt, Ni, and Fe/W, respectively. See ref. 16 for details.
[b] From Ref. 25 with corrections and additions specified below in footnotes e-i.
[c] Equations 6.41, 6.53, and 6.52, 6.54.
[d] Normalized for the stoichiometry C_2H_8. For C_2H_x (or $CH_y + CH_{x-y}$) the remaining $(8 - x)$ atoms H are assumed to be atomically adsorbed.
[e] For the average ($^{1.3}$A) state of CH_3CH, for which $D_{CH_2CH_2} - D_{CH_3CH} = 72$.
[f] For the ground (^{2}A) state of CH_3C, for which $D_{CH_2CH} - D_{CH_3C} = 45$.
[g] From $D_{CH_2CH_2} - D_{CH_3CH=1}$17.
[h] For the ground (^{1}A) state CH_2C, for which $D_{CHCH} - D_{CH_2C} = 44$.
[i] From $D_{CHCH} - D_{CHC} = 133$.

thus the total energy associated with the $-CH_2$ group is:

$$D_{(CH_2)_2} = D_{CH_2\cdot} + D_{C=C} = 183 + 172 = 355\,\text{kcal mole}^{-1}. \qquad (2)$$

If ethylene adsorption on Pt(111) is considered, then

$$Q_{C_2H_4} = Q_{(CH_2)_2} = \frac{(9/2)(0.6)^2(150)^2}{3(0.6)(150) + 8(355)} = 12\,\text{kcal mole}^{-1} \qquad (3)$$

In contrast, the newer UBI-QEP formalism allows the calculation of binding energies for polyatomic molecules without bond-energy partitioning. This capability is especially applicable to symmetric polyatomic molecules, such as ethylene, acetylene and hydrazine, but can also be applied to other polyatomic molecules such as nitrous oxide [20,22]. This approach gives a value of $11.0\,\text{kcal mole}^{-1}$ for the heat of chemisorption of C_2H_4 on Pt(111) [20], which is very close to the value of $12\,\text{kcal mole}^{-1}$ previously calculated.

6.3.4 Activation Barriers for Dissociation and Recombination on Metal Surfaces

Equation 6.40 can also be used for the BOC-MP/UBI-QEP treatment of dissociation of an AB molecule. In a traditional, one-dimensional Lennard-Jones (LJ) potential energy diagram, such as in Figure 5.1(b) or Figure 6.3 (a), the activation energy for dissociation refers to the intersection point of the molecular AB and atomic A + B curves, i.e., to $\Delta E_{AB,g}^{*LJ}$ in Figure 6.3(a). The transition state at this point would correspond to a configurational switch where $x_{AB}^{TS} = 0$, thus reducing equation 6.40 to

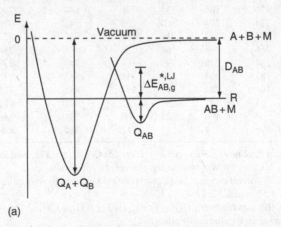

(a)

FIGURE 6.3. Chemisorption and dissociation of a diatomic AB molecule. (a) The traditional Lennard-Jones one-dimensional potential energy diagram of E vs. R, where R is the reaction coordinate.

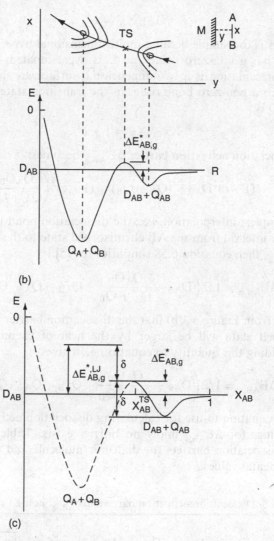

FIGURE 6.3. *continued* (b) The conventional two-dimensional potential energy diagram of E vs. R (x,y). The reaction coordinates are the A – B distance (x) and the AB – surface distance (y).The energy minima correspond to the (nondissociated) molecular chemisorbed state, $D_{AB} + Q_{AB}$, and the atomic (dissociated) chemisorbed state, $Q_A + Q_B$. The maximum energy is that of the transition state (TS) with some finite A – B bond length. (c) The multidimensional BOC potential energy diagram, similar to (b), but the reaction coordinate is now the A – B bond order, x_{AB}. The $M_{n'}$-AB bond order is conserved to unity ($x_A + x_{AB} + x_B = 1$) up to the transition state where $1 > x_{AB}^{TS} = c > 0$ and $\delta = (1/2)\left(\Delta E_{AB,g}^{*,LJ} + Q_{AB}\right)$. See ref. 16 for details. (Reprinted from ref. 16, copyright © 1990, with permission from Elsevier)

$$X_A + X_B = 1 \qquad (6.56)$$

However, this is too simple because multi-dimensional hypersurfaces occur; in fact, X_{AB}^{TS} has a non-zero value and E is overestimated. The minimally adequate representation is a two-dimensional surface, as shown in Figure 6.3(b), and for a non-zero bond order in the transition state, $x_{AB}^{TS} = c$, thus equation 6.40 is:

$$X_A + X_B = 1 - c \qquad (6.57)$$

and the dissociation activation barrier $\Delta E_{AB,g}^*$ becomes:

$$\Delta E_{AB,g}^* = (1 - c)^2 D_{AB} - (Q_A + Q_B) + (1 + c)^2 \left(\frac{Q_A Q_B}{Q_A + Q_B} \right) \qquad (6.58)$$

Using the simplest interpolation, i.e., the dissociation point is in the middle of the energy interval from the AB chemisorbed state to the LJ intersection point, $\Delta E_{AB,g}^{*LJ}$, then equation 6.58 simplifies to [15]:

$$\Delta E_{AB,g}^* = 1/2 \left(D_{AB} + \frac{Q_A Q_B}{Q_A + Q_B} - Q_{AB} - Q_A - Q_B \right) \qquad (6.59)$$

It is obvious from Figure 6.3(b) that the dissociation barrier associated with a chemisorbed state will be larger by the heat of chemisorption, Q_{AB}, therefore, adding this quantity to equation 6.58 gives

$$\Delta E_{AB,s}^* = 1/2 \left(D_{AB} + \frac{Q_A Q_B}{Q_A + Q_B} + Q_{AB} - Q_A - Q_B \right) \qquad (6.60)$$

which is the equation to use for calculating dissociation activation barriers. Negative values for $\Delta E_{AB,s}^*$ imply no barrier exists. Table 6.8 lists some calculated dissociation barriers for diatomic molecule and compares them with experimental values.

TABLE 6.8. Dissociation activation barriers $\Delta E_{AB,\,g}^*$: non-LJ corrections[a]

AB	Surface	D_{AB}	Q_A	Q_B	Q_{AB}	c^b	$\Delta E_{AB,\,g}^*$ Eq. 6.59	$\Delta E_{AB,\,g}^*$ Exp.
H_2	Fe(111)	104	62	62	7	0.08	2	0
	Ni(111)	104	63	63	7	0.06	1.5	2
	Cu(100)	104	56	56	5	0.09	7.5	5
N_2	Fe(110)	226	138	138	8	0.04	6.5	8
	Fe(100)	226	140	140	8	0.05	4.5	2.5
	Fe(111)	226	139	139	8	0.06	5	~ 0
CO	Ni(111)	257	171	115	27	0.10–0.12	6.5	—
	Ni(100)	257	171	130	30	0.09–0.11	~ 0	$-6, -7$
	W(110)	257	200	125	21	0.04–0.07	-6	-15
	Fe(111)	257	(200)	(125)	32	0.04–0.07	-12	-12
	Mo(100)	257	(200)	(125)	16	0.04–0.07	-4	-2

[a] See ref. 16 for details. All energies in kcal mole^{-1}.
[b] From equation 6.58.

For the reverse reaction representing the recombination of chemisorbed A and B, the LJ activation barrier is just [15]:

$$\Delta E_{rec} = \Delta E_{A-B}^{*LJ} = \frac{Q_A Q_B}{Q_A + Q_B} \tag{6.61}$$

However, it is obvious that this activation barrier for recombination cannot be less than the enthalpy change

$$\Delta H_{rec} = \Delta H_{AB} - \Delta H_{A+B} = -(D_{AB} + Q_{AB}) + (Q_A + Q_B) \tag{6.62}$$

going from the reactant AB_s on the surface (with $-\Delta H_{AB} = D_{AB} + Q_{AB}$) to the products A_s and B_s on the surface (with $-\Delta H_{A+B} = Q_A + Q_B$). Therefore, if equation 6.61 is less than equation 6.62, it can be assumed that the recombination barrier is equal to the enthalpy change for this step, ΔH_{rec}. More recent discussions about the use of the BOC-MP/UBI-QEP method to calculate activation barriers for surface diffusion and reactions on metal surfaces are provided elsewhere [19,21,22].

6.4   Use of a Rate Determining Step (RDS) and/or a Most Abundant Reaction Intermediate (MARI)

Boudart has discussed in detail the fact that the rate law derived from a complex catalytic cycle comprised of a number of elementary steps can frequently be represented by only two kinetically significant steps if the assumptions of a RDS and a MARI are invoked; however, ambiguities can develop which prevent one from distinguishing among different reaction models [11,26]. In similar fashion, but with perhaps less dramatic results, a L-H-type or H-W-type model [27] invoking more than one elementary surface reaction step can be greatly simplified by the presence of quasi-equilibrated steps which precede the RDS or, if a RDS does not exist, the series of slow steps on the surface. The SSA may also be required in the latter case to eliminate all unknown surface reaction intermediates from the rate law. Significant simplification is achieved with the assumption of a RDS.

Additional simplification is acquired when assumptions are made regarding the surface species to be included in the site balance. Frequently it is neither possible nor necessary to include every intermediate because many concentrations are going to be very small compared to others. By selecting those expected to dominate based upon the behavior of the rate and any available spectroscopic information, the number of important unknown surface coverages can be significantly decreased. At the limit, if a MARI exists, then only one surface species becomes important in the final rate derivation.

6.5   Evaluation of Parameter Consistency in Rate Expressions for Ideal Surfaces

It is certainly fair to ask why the assumption of an ideal surface should be used in modeling heterogeneous catalytic reactions when there is overwhelming evidence that real surfaces are nonideal in most cases. This is an implicit constraint when Langmuir isotherms are incorporated into the rate expression to give equations of the form $r = kP_A/(1 + \sum_i K_i P_i)$ and $r = kP_A P_B/(1 + \sum_i K_i P_i)^2$ where A and B are reactants and K_i represents the adsorption equilibrium constant for compound i, as discussed in detail in Chapter 7. A wide variety of these types of expressions can be seen in Table 7.10. Regardless, Langmuir-Hinshelwood (L-H) rate laws based on an ideal, or Langmuirian, surface usually can do a very satisfactory job of correlating the data. This paradox was recognized decades ago and, although it can be argued that L-H rate expressions are not too sensitive to the reaction model proposed, with similar rate laws being obtained from different models, the earliest rationale is probably still the best explanation, i.e., that on a nonideal (or nonuniform) surface the observed kinetic behavior will be dominated by the reaction occurring on the most active sites. As a consequence, the surface can appear to be ideal. The first treatment of the influence of surface nonuniformity on catalytic rates by Constable indeed resulted in this conclusion [28]. Subsequent treatments were also consistent with this conclusion, and a more recent analysis by Temkin [29,30] has shown this nicely in a quantitative manner, and this representation of a nonuniform surface will be discussed later in Chapter 8 based on the work of Temkin, as described by Boudart [9,11]. This is one reason why adsorption sites measured by chemisorption at different conditions may not correspond on a one-to-one basis with the active sites controlling the reaction under another set of conditions; however, this does not dispel the need to obtain TOFs, i.e., to determine and present normalized specific activities. Thus, if a group of sites with similar or identical properties dominate the overall reaction and they constitute only a fraction of the total sites but are widely distributed on the catalytic surface, the observed behavior may conform relatively well to that expected for a Langmuirian surface. Another argument supporting the utilization of ideal surfaces in the Langmuirian sense is the ease in which multiple adsorbates can be incorporated into the Langmuir isotherm to allow competitive adsorption. An examination of the derivations of the Freundlich and Temkin isotherms in Chapter 5 reveals that incorporation of numerous adsorbates into these expressions is not simple.

However, having acknowledged this paradox, there exists a set of rules and guidelines that allows assessment of the validity of the application of Langmuirian kinetics, i.e., the assumption of an ideal surface, to a particular set of kinetic data [26,31,32]. Criteria to evaluate whether rate parameters, such as the adsorption equilibrium constants appearing in the denominator of the Langmuir isotherm (and subsequently in the denominator of the rate expression, as we shall see), are consistent and physically meaningful are based upon the thermodynamic parameters contained in these constants (see Chapter 5, equation 5.38). Their utilization to evaluate the enthalpy and entropy of adsorption can allow one to discriminate among different rate expressions, all of which may give a similar statistical fit to the experimental data, and to reject some models in certain circumstances. Their application will be discussed in Chapter 7.

These criteria are listed in Table 6.9, and they are comprised of three strong rules and two guidelines. First, adsorption is invariably exothermic, thus the enthalpy of adsorption is negative, i.e., $-\Delta H_{ad}^o > 0 (Q_{ad} > 0)$. Second, there must be a decrease in entropy after adsorption, including dissociative adsorption of a diatomic molecule [32], thus $\Delta S_{ad}^o = S_{ad}^o - S_g^o < 0$, where S_g^o is the standard total entropy in the gas phase. Third, a molecule or atom cannot lose more entropy than it possesses prior to adsorption, thus $|-\Delta S_{ad}^o| < S_g^o$. Two less rigid guidelines can be proposed. First, an adsorbing species must lose at least one degree of translational freedom (if it becomes a 2-dimensional gas on the surface), and a calculation for a typical gas at a reasonable temperature gives a minimum value of approximately -10 entropy units (1 e.u. $= 1$ cal mole^{-1}K^{-1}) for this loss. Second, an empirical correlation reported by Everett [33] between the entropy and enthalpy of adsorption on carbons appeared to establish a linear upper limit on values associated with chemisorption [31], hence the suggested upper limit for the Guidelines in Table 6.9. An absolute value of $|-\Delta S_{ad}^o|$ falling well above this line should be viewed with suspicion. Additional criteria for evaluating kinetic rate parameters in specific elementary processes have been cited by Boudart [26] and these are listed in Table 6.10. It is important that the correct units be utilized when evaluating these rate constants.

The point of the preceding discussion is that the derivation and use of Langmuirian rate expressions can be justified. L-H models provide the catalytic investigator a convenient means to consider a reaction model, to derive a rate expression, to fit it easily to the experimental data using one of a number of commercially available software packages for optimization of the rate parameters and, finally, to evaluate the validity of the fitting parameters obtained. If all is consistent, one has a rate equation not only valid over the range of temperatures and pressures (or concentrations) utilized, but also with some predictive capacity outside of this range due to

TABLE 6.9. Criteria to evaluate adsorption equilibrium constants obtained as fitting parameters in a Langmuirian rate expression [from refs. 26,31,32]

Rule 1:	$-\Delta H_{ad}^o > 0 \, (Q_{ad} > 0)$
Rules 2 & 3:	$0 < -\Delta S_{ad}^o < S_g^o$
Guidelines:	$10 \stackrel{\sim}{<} -\Delta S_{ad}^o \stackrel{\sim}{<} 12.2 - 0.0014 \, \Delta H_{ad}^o$ (in cal mole^{-1})

where S_g^o is the standard entropy of the gas (1 atm).

TABLE 6.10. Criteria to evaluation rate constants for elementary steps on a surface (from Ref. 26, copyright © 1972 AIChE, reproduced with permission of the American Institute of Chemical Engineers)

1:	Adsorption

$$r_{ad} = Lk_a[A] = LA_a e^{-E_a/RT}[A]$$

Guideline: $LA_a \leq 10^4 \text{ cm s}^{-1}$

2:	Unimolecular Surface Reaction or Desorption

$$r_d = k_d[A\text{-}S] = LA_d e^{-E_d/RT}\theta_A$$

Guideline: $LA_d \leq 10^{28}$ site (or molecule) $\text{cm}^{-2}\,\text{s}^{-1}$
or, alternatively per site, $A_d \leq 10^{13}\,\text{s}^{-1}$

3:	Bimolecular Surface Reaction or Desorption

$$r_b = Lk_b'[A-S][A-S] = k_b[A-S]^2 = A_b e^{-E_b/RT}[A-S]^2 = k_b L^2 \theta_A^2$$

Guideline: $A_b \leq 10^{-4}\,\text{cm}^2\,\text{s}^{-1}$ molecule^{-1}

the insight acquired by the existence of a reaction model comprised of a series of elementary steps and reasonable approximations. Consequently, tests of the model pertaining to predictions based on the rate parameters and/or the MARI species can be proposed and conducted to further validate it. These include in situ characterization such as the use of infrared or Raman spectroscopy, for example. In the next chapter, various approaches to derive rate expressions on ideal (Langmuirian) surfaces will be examined. In fact, Mezaki and Inoue have compiled rate equations for almost 1000 reactions on heterogeneous catalysts, and all but two or three followed L-H (or Rideal-Eley) rate expressions which assumed an ideal surface [34].

References

1. R. P. Danner and T. E. Daubert. "Manual for Predicting Chemical Process Design Data", Design Institute for Physical Property Data, AIChE, NY, 1983.
2. I. Langmuir, *Trans. Faraday Soc.* 17 (1921) 621.
3. C. N. Hinshelwood, "Kinetics of Chemical Change in Gaseous Systems", Clarendon, Oxford, 1926.
4. C. N. Hinshelwood, "The Kinetics of Chemical Change", 4th Ed., Oxford Press, Oxford, 1940.
5. H. Eyring, *J. Phys. Chem.* 3 (1935) 107.
6. M. G. Evans and M. Polanyi, *Trans. Faraday Soc.* 31 (1935) 875.
7. K. J. Laidler, "Chemical Kinetics", 3rd Ed., Harper & Row, NY, 1987.
8. T. L. Hill, "An Introduction to Statistical Thermodynamics", Addison-Wesley, Reading, MA, 1960.
9. M. Boudart, "Kinetics of Catalytic Processes", Prentice-Hall, Englewood Cliffs, NJ, 1968.
10. M. Bodenstein, *Z. Phys. Chem.* 13 (1894) 56; 22 (1897) 1, 23; 29 (1899) 295.
11. M. Boudart and G. Djéga-Mariadassou, "Kinetics of Heterogeneous Catalytic Reactions", Princeton Press, Princeton, 1984.
12. S. W. Benson, "The Foundation of Chemical Kinetics", McGraw-Hill, NY, 1960.

13. C. G. Hill, Jr., "An Introduction to Chemical Engineering Kinetics and Reactor Design", Wiley, NY, 1977.
14. D. D. Eley, *Disc. Faraday Soc.* 8 (1950) 34.
15. D. P. Stevenson, *J. Chem. Phys.* 23 (1955) 203.
16. E. Shustorovich, *Adv. Catal.* 37 (1990) 101.
17. E. Shustorovich, *Surf. Sci. Rep.* 6 (1986) 1.
18. E. Shustorovich, *Acc. Chem. Res.* 21 (1988) 183.
19. E. Shustorovich and H. Sellers, *Surf. Sci. Rep.* 31 (1998) 1.
20. E. Shustorovich and A. V. Zeigarnik, *Surf. Sci.* 527 (2003) 137.
21. S. P. Baranov, L. A. Abramova, A. V. Zeigarnik and E. Shustorovich, *Surf. Sci.* 555 (2004) 20.
22. E. Shustorovich and A. V. Zeigarnik, To be published in "Kinetics and Catalysis", 2005 (Eng. translation).
23. L. Pauling, *J. Am. Chem. Soc.* 69 (1947) 542.
24. P. M. Morse, *Phys. Rev.* 34 (1929) 57.
25. "CRC Handbook of Chemistry and Physics", D. R. Lide, ed., CRC Press, Boca Raton, FL, 1990.
26. M. Boudart, *AIChE J.* 18 (1972) 465.
27. O. A. Hougan and K. M. Watson, "Chemical Process Principles, Part 3. Kinetics and Catalysis", Wiley, NY, 1947.
28. F. H. Constable, *Proc. Royal Soc. London A*, 108 (1925) 355.
29. M. I. Temkin, *Zhur. Fiz. Khim.* 31 (1957) 1.
30. M. I. Temkin, *Adv. Catal. Relat. Subj.* 28 (1979) 173.
31. M. Boudart, D. E. Mears, and M. A. Vannice, *Ind. Chim. Belge,* Special issue, 32 (1967) 281.
32. M. A. Vannice, S. H. Hyun, B. Kalpakci, and W. C. Liauh, *J. Catal.* 56 (1979) 358.
33. D. H. Everett, *Trans Faraday Soc.* 46 (1950) 957.
34. R. Mezaki and H. Inoue, "Rate Equations of Solid-Catalyzed Reactions", U. of Tokyo Press, Tokyo, 1991.
35. B. Sen and M. A. Vannice, *J. Catal.* 113 (1988) 52.
36. O. Kircher and O. A. Hougen, *AICHE J.* 3 (1957) 331.

Problem 6.1

The following first-order rate constants were obtained for the thermal decomposition of ethane:

Calculate the activation energy and the pre-exponential factor.

Temperature (°C)	Rate constant ($s^{-1} \times 10^5$)
550	1.3
560	2.3
570	4.1
580	6.2
590	11.5
600	17.7
610	28.6
620	46.2
630	70.8

138   6. Kinetic Data Analysis and Evaluation of Model Parameters

Problem 6.2

Using statistical thermodynamics, determine the temperature dependence of the pre-exponential factor for:

a) the gas-phase reaction of two N atoms to form N_2.
b) the bimolecular gas-phase reaction of $H_2O + CO$ to form H_2 and CO_2.
c) the bimolecular gas-phase reaction of $H_2 + Cl_2$ to form HCl.

Problem 6.3

Use the steady-state approximation and derive the rate expression for the following sequence of elementary steps describing a chain reaction. Define your rate clearly.

$$Br_2 \xrightarrow{k_i} 2\ Br\cdot \qquad \text{Initiation}$$

$$Br\cdot + H_2 \underset{k_{-1}}{\overset{k_1}{\rightleftarrows}} HBr + H\cdot$$

$$H\cdot + Br_2 \xrightarrow{k_2} HBr + Br\cdot \qquad \Big\} \text{ Propagation}$$

$$Br\cdot + Br\cdot \xrightarrow{k_t} Br_2 \qquad \text{Termination}$$

$$\overline{Br_2 + H_2 \implies 2\ HBr}$$

Problem 6.4

The series of elementary steps below has been proposed to describe acetaldehyde pyrolysis to methane and CO. Derive the steady-state rate expression making the usual long-chain approximation; i.e., that the "chain length" is very high. What comprises the apparent activation energy?

$$CH_3CHO \xrightarrow{k_1} CH_3\cdot + CHO\cdot \qquad \text{Initiation}$$

$$CH_3\cdot + CH_3CHO \xrightarrow{k_2} CH_3CO\cdot + CH_4$$

$$CH_3CO\cdot \xrightarrow{k_3} CH_3\cdot + CO \qquad \Big\} \text{ Propagation}$$

$$2\ CH_3\cdot \xrightarrow{k_4} C_2H_6$$

$$CHO\cdot + wall \xrightarrow{k_5} coke \qquad \Big\} \text{ Termination}$$

$$\overline{CH_3CHO \implies CH_4 + CO}$$

Problem 6.5

Using the BOC-MP method, calculate the heat of adsorption for the following four species. All the information and values needed are contained in the various tables in section 6.3 of this chapter. State the appropriate adsorption mode, i.e., $(\eta^x \mu_y)$.

(a) Nitric oxide on Fe(110) – what end of NO should bond to Fe? What is the preferred coordination, i.e., is $y = 1$ or 2? Why?

(b) H_2O on Pd(100)

(c) Acetylene ($HC \equiv CH$) on Pt(111)

(d) The methylene radical ($\cdot CH_2$) on Pt(100)

Problem 6.6

Based on the BOC-MP method, calculate the heat of adsorption for the following species. All information and values needed are contained in the various tables in Section 6.3 of this chapter.

(a) NO on Ni(100) $(\eta^1 \mu_2)$

(b) C_2H_4 on Pd(111) $(\eta^2 \mu_2)$

(c) NH_3 on Fe(110) $(\eta^1 \mu_1)$ (Check ref. 25 for D_{AB} value)

(d) NH_2 and NH on Fe(110) (Check ref. 25 for D_{AB} value)

Problem 6.7

The kinetics of acetone hydrogenation on Pt catalysts have been studied by Sen and Vannice (See Problem 7.9) [35]. To determine which bonding mode is more probable, the heat of adsorption for an on-top $\eta^1 \mu_1$ adsorbed acetone species was compared to that for a di-σ-bonded $\eta^2 \mu_2$ species which had a C atom and the O atom interacting with a close-packed Pt(111) surface. What are these two values? Which species is favored?

Problem 6.8

Derive Guideline 3 in Table 6.10 for a monatomic species of 30 amu:

a) Using collision theory and assuming a 2-dimensional gas.

b) Using absolute rate theory and assuming immobile adsorption.

c) Using absolute rate theory and assuming a 2-dimensional gas.

Problem 6.9

Kircher and Hougen studied NO oxidation, $2NO + O_2 \longrightarrow 2NO_2$, over activated carbon and SiO_2 in the presence of water and proposed a Rideal-Eley mechanism between O_2 and $(N_2O_2)_{ad}$, which gave a rate expression of:

$$r = \frac{kP_{NO}^2 P_{O_2}}{\left(1 + K'P_{NO}^2 + K_{NO_2}P_{NO_2} + K_{H_2O}P_{H_2O}\right)}$$

The values below were reported [36]. Properly write the elementary steps implied. Does the reaction sequence suggested by this equation appear to be reasonable? Why?

	Temp.	303 K	318 K	333 K
Act. C				
	k	3,500	5,300	5,500
	$K_{NO_2}(atm^{-1})$	23	35	75
	$K_{H_2O}(atm^{-1})$	287	339	301
SiO$_2$				
	k	251	114	63
	$K_{NO_2}(atm^{-1})$	61	20	11
	$K_{H_2O}(atm^{-1})$	1020	206	54

7
Modeling Reactions on Uniform (Ideal) Surfaces

The concept of a rate determining step (RDS) has been previously discussed and defined in Chapter 2. Numerous kinetic studies have had their experimental data satisfactorily correlated by a rate expression derived from a reaction sequence invoking a RDS on an ideal surface, and the approaches associated with this assumption are typically referred to as Langmuir-Hinshelwood (L-H)-type or Hougan-Watson (H-W)-type [1] models. The former group assumes a reaction on the surface governs the rate, with one of the elementary steps in the reaction sequence constituting a RDS, thus all the adsorption/desorption steps are quasi-equilibrated and a Langmuir isotherm can be used to relate surface concentrations to bulk concentrations or partial pressures. Consequently, these L-H models are subject to the same assumptions associated with the Langmuir isotherm (see Chapter 5.3), but with the awareness that the active sites described by the site balance in the Langmuir isotherm may not be identical to the number of sites measured by adsorption under nonreacting conditions. The latter (H-W) group is broader because it also allows for an adsorption step or a desorption step to be a RDS, but Langmuir isotherms are again used, when appropriate, to relate surface and bulk concentrations. Of course, it is possible that no RDS exists in the reaction mechanism, and this situation, which necessitates the use of the SSA, will also be examined.

The paradox of using ideal surfaces to represent surfaces known to be nonideal or nonuniform was addressed in the preceding chapter, and justification was provided for doing so. Thus let us proceed with some of the various approaches for proposing reaction models and deriving rate expressions. Some of the oldest and most straightforward approaches involve the use of L-H-type or H-W-type reaction models, which invoke a RDS [1]. We will start with this family of sequences by examining the simplest surface reactions first, and then reaction models without a RDS will be discussed.

7.1 Reaction Models with a RDS – Unimolecular Surface Reactions

The simplest reaction model can be depicted by the following sequence of elementary steps which describes a reversible isomerization reaction of A to B, where S is an active site:

$$(7.1) \qquad A + S \; \overset{K_A}{\rightleftharpoons} \; A - S \qquad (QE)$$

$$(7.2) \qquad A - S \; \underset{\overleftarrow{k}}{\overset{\overrightarrow{k}}{\rightleftharpoons}} \; B - S \qquad (RDS)$$

$$(7.3) \qquad B - S \; \overset{K'_B}{\rightleftharpoons} \; B + S \qquad (QE)$$

$$\overline{\qquad A \; \overset{\longrightarrow}{\rightleftharpoons} \; B \qquad}$$

The quasi-equilibrated (QE) steps and the RDS step are designated. This reaction model represents a series of elementary steps comprising a closed sequence and a catalytic cycle. In its reversible form, from the law of mass action the net rate is (Chapter 2.7):

$$-\frac{d[A]}{dt} = r = \overrightarrow{r} - \overleftarrow{r} = \overrightarrow{k}[A - S] - \overleftarrow{k}[B - S] \qquad (7.1)$$

where [A–S] and [B–S] now represent surface concentrations (molecule cm^{-2}). Then, from Chapter 5 we see that $K'_B = 1/K_B$, where K_A and K_B are adsorption equilibrium constants, with:

$$[A - S] = L\theta_A = LK_A P_A/(1 + K_A P_A + K_B P_B) \qquad (7.2)$$

and

$$[B - S] = L\theta_B = LK_B P_B/(1 + K_A P_A + K_B P_B) \qquad (7.3)$$

after using the site balance of:

$$L = [S] + [A - S] + [B - S] \qquad (7.4)$$

where L is the total concentration of active sites or, alternatively:

$$1 = \theta_v + \theta_A + \theta_B \qquad (7.5)$$

Substitution of equations 7.2 and 7.3 into equation 7.1 gives the most general form of this rate expression:

$$r = \frac{L(\overrightarrow{k}K_A P_A - \overleftarrow{k}K_B P_B)}{(1 + K_A P_A + K_B P_B)} \qquad (7.6)$$

If step 7.2 is irreversible ($\vec{k} \gg \overleftarrow{k}$ for example), the second term in the numerator disappears and, further, if [B–S] is very low because P_B is very low (as in a differential reactor) or K_B is very small, then [A–S] is the MARI and the last term in equation 7.4 can be neglected. The rate then simplifies to:

$$r = \frac{L\vec{k}K_A P_A}{1 + K_A P_A} = \frac{k_{ap}P_A}{1 + K_A P_A} \qquad (7.7)$$

where the apparent rate constant, k_{ap}, represents a combination of terms. Figure 7.1a illustrates the dependence of the rate on the partial pressure of the reactant. Plots of a linearized form of the rate law such as:

$$\frac{1}{r} = \left(\frac{1}{L\vec{k}K_A}\right)\frac{1}{P_A} + \frac{1}{L\vec{k}} \qquad (7.8)$$

obtained at different temperatures can be used to extract values for $L\vec{k}$ and K_A. Arrhenius plots of these two fitting parameters provide values for the pre-

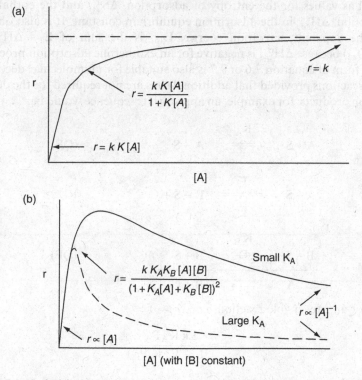

(a)

$r = k$

$r = \dfrac{k\, K\,[A]}{1 + K\,[A]}$

$r = k\, K\,[A]$

[A]

(b)

$r = \dfrac{k\, K_A K_B\,[A]\,[B]}{(1 + K_A[A] + K_B\,[B])^2}$

Small K_A

Large K_A

$r \propto [A]^{-1}$

$r \propto [A]$

[A] (with [B] constant)

FIGURE 7.1. Variation of rate, r, with concentration, [A], for different types of surface reactions: (a) single-site unimolecular process; (b) dual-site (bimolecular) process involving A and B via a Langmuir-Hinshelwood mechanism assuming one type of site;

Continued

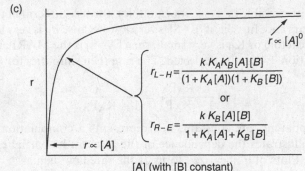

$$r_{L-H} = \frac{k\,K_A K_B [A][B]}{(1+K_A[A])(1+K_B[B])}$$

or

$$r_{R-E} = \frac{k\,K_B[A][B]}{1+K_A[A]+K_B[B]}$$

FIGURE 7.1. *continued* (c) dual-site (bimolecular) process involving A and B via either a Langmuir-Hinshelwood mechanism assuming two types of site or a Rideal-Eley mechanism in which A may also adsorb but adsorbed A is unreactive.

exponential factor, A_o, and the activation energy, E_a, in the rate constant $L\vec{k}$ as well as values for the entropy of adsorption, ΔS^o_{ad}, and the enthalpy of adsorption, ΔH^o_{ad}, in the adsorption equilibrium constant. It is also obvious that the apparent activation energy, E_{ap}, is the sum of $E_a + \Delta H^o_{ad}$ (or $E_a - Q_{ad}$) because ΔH^o_{ad}, is negative for an exothermic adsorption process.

The form of equation 7.6 or 7.7 is also suitable for unimolecular decomposition reactions provided that additional sites are not required for the decomposition products; for example, an appropriate sequence would be:

$$
(7.4) \qquad A + S \;\overset{K_A}{\rightleftharpoons}\; A - S \qquad\qquad (QE)
$$

$$
(7.5) \qquad A - S \;\overset{\vec{k}}{\underset{\overleftarrow{k}}{\rightleftharpoons}}\; B - S + C \qquad\qquad (RDS)
$$

$$
(7.6) \qquad B - S \;\overset{K'_B}{\rightleftharpoons}\; B + S \qquad\qquad (QE)
$$

$$
\overline{\qquad A \quad\overset{}{\rightleftharpoons}\quad B + C \qquad}
$$

and, for an irreversible reaction, i.e., $\vec{r} \gg \overleftarrow{r}$:

$$
r = \frac{L\vec{k}\,K_A P_A}{1 + \sum_i K_i P_i} \tag{7.9}
$$

$$
\text{where } [S] = L \Big/ \Big(1 + \sum_i K_i P_i\Big) \text{ and thus } \theta_v = 1 \Big/ \Big(1 + \sum_i K_i P_i\Big) \tag{7.10}
$$

Note that C is not adsorbed, so it would not be included in the site balance with this sequence of steps.

However, if in a unimolecular decomposition reaction at least one additional active site is required in the RDS (or any step preceding it), then a different rate expression is obtained. Consider again A decomposing to B + C, but now:

$$(7.7) \qquad A + S \overset{K_A}{\rightleftharpoons} A - S \qquad (QE)$$

$$(7.8) \qquad A - S + S \overset{\overrightarrow{k}}{\underset{\overleftarrow{k}}{\rightleftharpoons}} B - S + C - S \qquad (RDS)$$

$$(7.9) \qquad B - S \overset{K'_B}{\rightleftharpoons} B + S \qquad (QE)$$

$$(7.10) \qquad \frac{C - S \overset{K'_C}{\rightleftharpoons} C + S}{A \rightleftharpoons B + C} \qquad (QE)$$

Where K'_B and K'_C are again reciprocals of the adsorption equilibrium constants. The reverse reaction is obviously bimolecular on the surface, and the rate of the forward RDS requires a vacant site and is therefore:

$$-\frac{d[A]}{dt} = \overrightarrow{r} = \overrightarrow{k}[A - S][S] = L\overrightarrow{k}\,\theta_A\theta_v \qquad (7.11)$$

Consequently, the denominator in the rate expression is now *squared* and the rate law from equation 7.11 becomes (for an irreversible reaction):

$$r = L\overrightarrow{k}K_AP_A/(1 + K_AP_A + K_BP_B + K_CP_C)^2 \qquad (7.12)$$

Note that substitution of equations 7.2 (including K_CP_C) and 7.10 into equation 7.11 does NOT result in a second-order dependence on L (i.e., L^2), as mistakenly stated in various texts and papers, because nearest neighbor sites are required in step 7.8 above, and the probability for site pairs is 1/2 Z/L[2], where Z is the number of nearest neighbor sites around a site on the surface. Consequently, the L^2 term must be replaced by 1/2 ZL, and the rate is *always* proportional to the site density L in the regime of kinetic control.

All three reaction sequences provided in this section have represented simple L-H models with a single elementary step on the surface representing the RDS. In many, if not most, cases the catalytic reaction on the surface may consist of a sequence of elementary steps, one or all of which may represent the slow step in the catalytic cycle. If it is just a single step, then this is the RDS and all other steps can be assumed to be quasi-equilibrated. If a series of two or more steps represents slow surface reactions, then there is no

RDS and the SSA may have to be used to determine surface concentrations for some species; however, all adsorption/desorption steps are still quasi-equilibrated and Langmuir isotherms can still be used to describe the surface concentrations of species associated with these steps, hence the term L-H-type reactions. These statements also apply to the bimolecular surface reactions discussed in the next section.

In either situation with single-site unimolecular reactions (steps 7.1-7.3 or 7.4-7.6), one can see that limiting cases for an irreversible reaction can exist for equation 7.7 depending on the strength of chemisorption, i.e., the relative surface coverages of A, B and C, and Laidler has compiled a list of older examples from the literature [3]. They include the following:

a) If chemisorption of A is very strong and A nearly saturates the surface, then $\theta_A \gg \theta_B$, θ_C, θ_v and equation 7.7 leads to $r = kP_A^0$ and the reaction is zero order in regard to A. This was found during ammonia decomposition on W between 973 and 1573 K.

b) If the surface coverage of all species is very low, then $\theta_v \gg \theta_A$, θ_B, θ_C and equation 7.7 gives $r = kP_A$, a 1st-order reaction in A. This was observed for phosphine decomposition on Mo near 900 K.

c) If chemisorption of B is more pronounced than that of A and C, then $\theta_B \gg \theta_A$, θ_C and product inhibition occurs, i.e., equation 7.7 becomes $r = kP_A/(1 + K_BP_B)$. This was observed for O_2 inhibition during nitrous oxide decomposition on Pt at about 900 K.

d) If chemisorption of B is especially strong so that it nearly saturates the surface, then $\theta_B \gg \theta_v$, θ_A, θ_C, and an inverse 1st-order dependence on B can occur, i.e., $r = kP_A/P_B$, as observed during ammonia decomposition over Pt between 1273 and 1773 K.

Some additional examples of single-site unimolecular reactions have been reported by Vannice and coworkers in laboratory studies of N_2O decomposition. At low conversions this reaction can be readily represented by:

$$(7.11) \qquad 2[N_2O + * \xrightarrow{K_{N_2O}} N_2O*] \qquad \text{(QE)}$$

$$(7.12) \qquad 2[N_2O* \xrightarrow{\vec{k}} N_2 + O*] \qquad \text{(RDS)}$$

$$(7.13) \qquad 2\,O* \xrightarrow{1/K_{O_2}} O_2 + 2\,* \qquad \text{(QE)}$$
$$\overline{\qquad\qquad 2\,N_2O \implies 2\,N_2 + O_2 \qquad\qquad}$$

where * represents an active site, step 7.12 is the RDS, and the stoichiometric number for steps 7.11 and 7.12 in the catalytic cycle is 2. The areal rate equation derived from this with a site balance of

$$L = [*] + [N_2O*] + [O*] \tag{7.13}$$

is:

$$r_a = \frac{1}{A}\frac{dN_{N_2}}{dt} = L\vec{k}\theta_{N_2O} = \frac{kK_{N_2O}P_{N_2O}}{\left(1 + K_{N_2O}P_{N_2O} + K_{O_2}^{1/2}P_{O_2}^{1/2}\right)} \tag{7.14}$$

The kinetics of this reaction over a 4% Sr/La_2O_3 catalyst were examined under differential reactor conditions in both the absence and the presence of dioxygen in the feed stream [4]. The fitting of the experimental data to equation 7.14 is shown in Figure 7.2, with P_{N_2O} in units of Torr. The influence of O_2 was negligible, which produced the following rate law at 923 K:

$$r(\mu mol\ N_2\ s^{-1}\ m^{-2}) = k_{ap}P_{N_2O}/(1 + K_{N_2O}P_{N_2O})$$
$$= 62\ P_{N_2O}/(1 + 21\ P_{N_2O}) \tag{7.15}$$

with an apparent activation energy of 24 kcal mole^{-1} for k_{ap}:

$$k_{ap} = kK_{N_2O} = A_{ap}e^{-E_{ap}/RT} = LA_de^{\Delta S^{\circ}_{N_2O}/R}e^{-E_{ap}/RT} \tag{7.16}$$

where A_d is the preexponential factor associated with the unimolecular decomposition of N_2O. The results indicate the coverage of sites by O atoms is very low and can be neglected in the site balance. Also, the representation of reaction 7.13 as being quasi-equilibrated is justified by the rapid isotopic

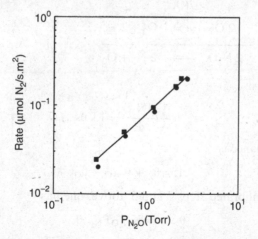

FIGURE 7.2. Rate dependence on N_2O pressure for N_2O decomposition on Sr/La_2O_3: (●) without O_2 in the feed; (■) with 1% O_2 in the feed, (——) fitting obtained with Eq. 7.14. (Reprinted from ref. 4, copyright © 1996, with permission from Elsevier)

exchange of O_2 on La_2O_3 at these temperatures [5]. A more detailed example of a single-site reaction is provided by Illustration 7.1.

Illustration 7.1 – N_2O Decomposition on Mn_2O_3: A L-H Reaction Model

Yamashita and Vannice have investigated N_2O decomposition on Mn_2O_3 in detail and, in addition, N_2O chemisorption was measured at temperatures between 273 and 353 K [6]. The Mn_2O_3 had a surface area of $31.8\,m^2\,g^{-1}$ after heating at 773 K in He. Rates were determined at five different temperatures from 598 to 648 K under differential reaction conditions (conv. $\leq$ 0.10) as a function of both N_2O and O_2 partial pressures. The Mn_2O_3 retained its stoichiometry under reaction conditions, as verified by both XRD measurements on the used catalyst and a constant N_2/O_2 ratio of 2 in the effluent stream during reaction. A Rideal-Eley mechanism in which the O_2 desorption step was replaced by the reaction $N_2O + O* \rightleftharpoons N_2 + O_2 + *$ was found to be inconsistent with the kinetic behavior [6]. Consequently, the proposed reaction sequence was again the L-H model previously described by steps 7.11 – 7.13, i.e.,

$$(1) \qquad 2[N_2O + * \underset{}{\overset{K_{N_2O}}{\rightleftharpoons}} N_2O*]$$

$$(2) \qquad 2[N_2O* \overset{\vec{k}}{\xrightarrow{A}} N_2 + O*] \qquad\qquad (RDS)$$

$$(3) \qquad 2\,O* \underset{}{\overset{1/K_{O_2}}{\rightleftharpoons}} O_2 + 2\,*$$

$$\overline{\qquad 2\,N_2O \implies 2\,N_2 + O_2 \qquad}$$

The areal rate is:

$$r_a = \frac{1}{A}\frac{dN_{N_2}}{dt} = L\vec{k}\,\theta_{N_2O} \qquad (1)$$

and the site balance is

$$1 = \theta_v + \theta_{N_2O} + \theta_O \qquad (2)$$

With quasi-equilibrated steps 1 and 3 above, one has

$$\theta_{N_2O} = K_{N_2O}P_{N_2O}\theta_v \qquad (3)$$

and

$$\theta_O = K_{O_2}^{1/2}P_{O_2}^{1/2}\theta_v \qquad (4)$$

When equations *3* and *4* are substituted first into equation *2* and then into equation *1*, the final rate expression is:

$$r_a = \frac{L\vec{k}\,K_{N_2O}P_{N_2O}}{\left(1 + K_{N_2O}P_{N_2O} + K_{O_2}^{1/2}P_{O_2}^{1/2}\right)} = \frac{k'K_{N_2O}P_{N_2O}}{\left(1 + K_{N_2O}P_{N_2O} + K_{O_2}^{1/2}P_{O_2}^{1/2}\right)} \tag{5}$$

which is identical to that proposed earlier by Rheaume and Parravarro for this reaction over Mn_2O_3 [7]. The results are provided in Figure 7.3, which also shows the optimum fitting obtained using equation *5* with the optimized fitting parameters listed in Table 7.1.

As an example, at 623 K with P units of atm, the rate expression is:

$$r(\mu mol\ s^{-1}\ m^{-2}) = 0.12\,P_{N_2O}\Big/\left(1 + 2.8\,P_{N_2O} + 12.6\,P_{O_2}^{1/2}\right) \tag{6}$$

Arrhenius plots of the parameters in Table 7.1 are very linear, as shown in Figure 7.4, and a value of 31 kcal mole^{-1} was obtained for $\vec{k}$ (in the RDS) while values of $\Delta H_{ad}^0 = -7$ kcal mole^{-1} and $\Delta S_{ad}^0 = -9$ e.u. (e.u. = entropy units = cal mole^{-1}K^{-1}) were calculated for N_2O adsorption, and values of $\Delta H_{ad}^0 = -22$ kcal mole^{-1} and $\Delta S_{ad}^0 = -26$ e.u. were obtained for O_2 chemisorption. These thermodynamic parameters satisfy all the criteria in

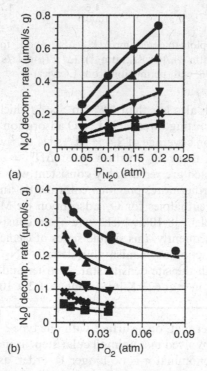

FIGURE 7.3. Fit between model and experimental data for N_2O decomposition on Mn_2O_3: (●) 648K, (▲) 638K, (▼) 623K, (✗) 608K, (■) 598K. (a) No oxygen in the feed, (b) $P_{N_2O} = 0.1$ atm. (Reprinted from ref. 6, copyright © 1996, with permission from Elsevier)

TABLE 7.1. Model Parameters (from Optimization) for N_2O Decomposition over Mn_2O_3 (Reprinted from ref. 6, copyright © 1996, with permission from Elsevier)

Fitting Parameter	Temperature				
	598K	608K	623K	638K	648K
$K_{N_2O}(atm^{-1})$	3.04	2.85	2.80	2.06	2.01
$k'(\mu mol\ s^{-1} m^{-2} \times 10^2)$	1.70	2.57	4.29	9.15	12.1
$K_{O_2}(atm^{-1})$	305	205	158	110	62

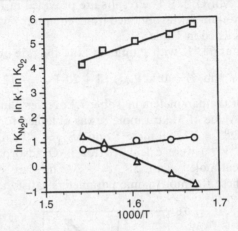

FIGURE 7.4. Arrhenius plots to determine ΔH_{ad}^o, ΔS_{ad}^o, and E' for N_2O decomposition on Mn_2O_3: (o) $K_{N_2O}(atm^{-1})$, (□) $K_{O_2}(atm^{-1})$, (△) $k'(\mu mole/s \cdot g)$. (Reprinted from ref. 6, copyright © 1996, with permission from Elsevier)

Table 6.9, and LA_d is also less than 10^{28} cm^{-2} s^{-1} (which satisfies Guideline 2, Table 6.10). After fitting three sets of N_2O adsorption data to a Langmuir isotherm, a saturation coverage of 2.4×10^{18} sites m^{-2} was determined, and Arrhenius plots of these K_{ad} values gave $\Delta H_{ad}^0 = -5\,kcal\ mole^{-1}$ and $\Delta S_{ad}^0 = -12$ e.u., which are remarkably consistent with the values from the kinetic analysis. Furthermore, previous calorimetric and kinetic investigations had provided enthalpies for O_2 adsorption on Mn_2O_3 ranging from -12 to $-24\,kcal\ mole^{-1}$ [8-10], which were again consistent with the value obtained here. Consequently, this simple series of elementary steps appears to be a consistent reaction mechanism that describes N_2O decomposition on Mn_2O_3. Finally, with the site density stated earlier and a measured rate of 4.8×10^{-3} $\mu mol\ s^{-1}\ m^{-2}$ at 623 K [6], a TOF = 1.2×10^{-3} s^{-1} is calculated.

It was stated earlier that, on a surface with one type of site, if more than a single active site is involved in a unimolecular step, as indicated by reactions 7.7-7.10, then the denominator is no longer 1st-order, as shown in equation 7.6, but it is raised to an appropriate power, such as that given in equation 7.12. A recent example of such a decomposition reaction is isopropyl alcohol dehydrogenation to produce acetone, and this is discussed in Illustration 7.2.

Illustration 7.2 – Isopropanol Dehydrogenation on Cu: A L-H Reaction Model

Isopropyl alcohol dehydrogenation over copper catalysts has been studied by Rioux and Vannice [11], and a detailed kinetic study was conducted of this reaction over a 0.98% Cu/carbon catalyst between 433 and 473 K and at a total pressure of 1 atm. The support was an activated carbon that had been given a high temperature treatment at 1223 K under H_2 to remove S impurities (1300 ppm), and it had a surface area of $1140\,m^2\,g^{-1}$. After a pretreatment in H_2 at 573 K, CO and N_2O adsorption showed that only metallic Cu (Cu^o) was present at the surface of the metal crystallites, in agreement with the XRD pattern, and the Cu dispersion was 0.11, indicating an average Cu^o crystallite size of about 9.8 nm.

Activity dependencies on isopropyl alcohol (IPA), H_2, and acetone (ACE) were determined at four temperatures, and they are shown in Figures 7.5–7.7, respectively. Fitting these data to a power rate law gave the reaction orders in

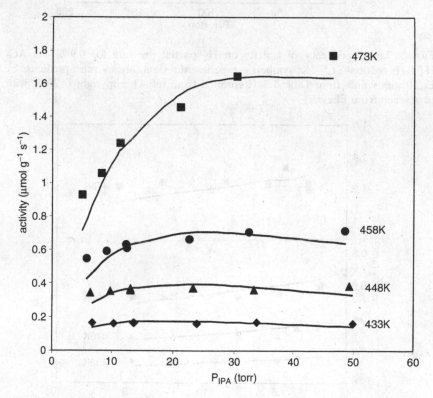

FIGURE 7.5. Dependency of activity on IPA (isopropyl alcohol) partial pressure for 0.98% Cu/AC-HTT-H_2 reduced at 573K; symbols – experimental data, lines – rates predicted by Eq. 9 using values from Table 7.3. (Reprinted from ref. 11, copyright © 2003, with permission from Elsevier)

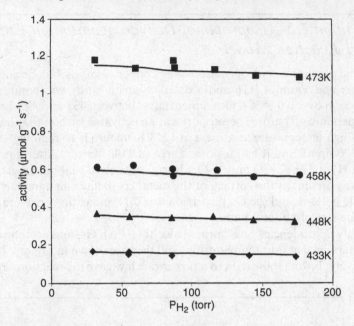

FIGURE 7.6. Dependency of activity on H_2 partial pressure for 0.98% Cu/AC-HTT-H_2 reduced at 573K; symbols – experimental data, lines – rates predicted by Eq. 9 with values from Table 7.3. (Reprinted from ref. 11, copyright © 2003, with permission from Elsevier)

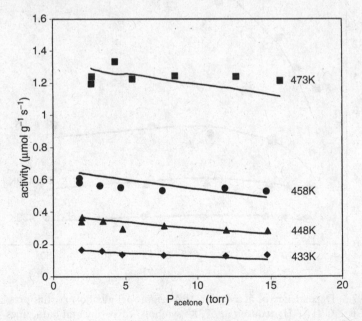

FIGURE 7.7. Dependency of activity on acetone partial pressure for 0.98% Cu/AC-HTT-H_2 reduced at 573K; symbols – experimental data, lines – rates predicted by Eq. 9 using values from Table 7.3. (Reprinted from ref. 11, copyright © 2003, with permission from Elsevier)

TABLE 7.2. Reaction Orders for Isopropyl Alcohol Dehydrogenation over 0.98% Cu/Activated Carbon[a] (Reprinted from ref. 11, copyright © 2003, with permission from Elsevier)

T_{RXN} (K)	Reaction Order		
	Isopropanol	H_2	Acetone
433	0	−0.11	−0.11
448	0.04	−0.07	−0.10
458	0.14	−0.05	−0.05
473	0.34	−0.04	0

[a]Reduced at 573 K, with (HTT-H_2) carbon sample

Table 7.2. Utilization of the Weisz parameter (see Chapter 4.2.3) gave values well below 0.01, thus verifying the absence of internal diffusional limitations, and this catalyst showed no deactivation after nearly 20 h on stream. The apparent activation energy, E_{ap}, over the temperature range examined was 20.5 kcal mole^{-1} under differential reactor conditions with only IPA in the feed, but adding H_2 to the feed increased E_{ap} by several kcal mole^{-1}.

A number of L-H-type and H-W-type reaction sequences were considered, rate expressions were derived, and their capability to fit the data was examined [12]. For example, the assumption of IPA adsorption as the RDS gave only a 1st-order dependence on IPA, in contradiction to Fig. 7.5, and the assumption that removal of the 2nd H atom was the RDS predicted a H_2 dependency between $-1/2$ and $-3/2$, which is inconsistent with Figure 7.6; consequently, both these models were discarded. The best reaction sequence to correlate the data, determined by quality of fit and the physical consistency of rate parameters based on Tables 6.9 and 6.10, was a L-H-type model requiring an additional vacant site in the RDS for decomposition, i.e.,

$$(1) \qquad (CH_3)_2 \, CHOH + * \quad \overset{K_{IPA}}{\rightleftharpoons} \quad (CH_3)_2 \, CHOH*$$

$$(2) \qquad (CH_3)_2 \, CHOH* + * \quad \overset{k_1}{\longrightarrow} \quad (CH_3)_2 \, CHO* + H* \qquad (RDS)$$

$$(3) \qquad (CH_3)_2 \, CHO* + * \quad \overset{k_2}{\rightleftharpoons} \quad (CH_3)_2 \, CO* + H*$$

$$(4) \qquad (CH_3)_2 \, CO* \quad \overset{1/K_{ACE}}{\rightleftharpoons} \quad (CH_3)_2 \, CO + *$$

$$(5) \qquad 2\,H* \quad \overset{1/K_{H_2}}{\rightleftharpoons} \quad H_2 + 2\,*$$

$$IPA \quad \Longrightarrow \quad ACE + H_2$$

Here the loss of the first H atom to form an isopropoxide species (ISO) on the surface, which is consistent with surface science studies by Friend and coworkers [13,14], represents a RDS. Consequently, the rate is:

$$r_m = -\frac{1}{m}\frac{dN_{IPA}}{dt} = Lk_1\theta_{IPA}\theta_v \;(\mu mole\; s^{-1}\; g\; cat^{-1}) \tag{1}$$

and the quasi-equilibrated adsorption/desorption processes (steps 1, 4 and 5) provide:

$$\theta_{IPA} = K_{IPA}P_{IPA}\theta_v \tag{2}$$

$$\theta_H = K_{H_2}^{1/2}P_{H_2}^{1/2}\theta_v \tag{3}$$

$$\theta_{ACE} = K_{ACE}P_{ACE}\theta_v \tag{4}$$

which can be substituted into the quasi-equilibrated step controlling the surface isopropoxide (ISO) coverage:

$$K_2 = \theta_{ACE}\theta_H/\theta_{ISO}\theta_v \tag{5}$$

to give:

$$\theta_{ISO} = K_{ACE}K_{H_2}^{1/2}K_2^{-1}P_{ACE}P_{H_2}^{1/2}\theta_v \tag{6}$$

These terms can now be placed in the site balance which, if no surface intermediate is omitted because in situ IR spectroscopy indicated the presence of an isopropoxide species [11], is:

$$1 = \theta_v + \theta_{IPA} + \theta_H + \theta_{ACE} + \theta_{ISO} \tag{7}$$

and this yields:

$$\theta_v = \left(1 + K_{IPA}P_{IPA} + K_{H_2}^{1/2}P_{H_2}^{1/2} + K_{ACE}P_{ACE} + K_{ACE}K_{H_2}^{1/2}K_2^{-1}P_{ACE}P_{H_2}^{1/2}\right)^{-1} \tag{8}$$

Finally, substituting first equation 2 and then equation 8 into equation 1 gives the rate equation:

$$r_m = Lk_1K_{IPA}P_{IPA}\theta_v^2$$

$$= \frac{kK_{IPA}P_{IPA}}{\left(1 + K_{IPA}P_{IPA} + K_{H_2}^{1/2}P_{H_2}^{1/2} + K_{ACE}P_{ACE} + K_{ACE}K_{H_2}^{1/2}K_2^{-1}P_{ACE}P_{H_2}^{1/2}\right)^2} \tag{9}$$

Using a commercial optimization software package, the fits to the data in Figures 7.5-7.7 were obtained. With five fitting parameters, one can validly argue that a very satisfactory fit should have been possible; however, as discussed previously, criteria are available to evaluate these rate parameters. The optimum parameters at four temperatures, given in Table 7.3, where $k = Lk_1$ and $K' = K_{ACE}K_{H_2}^{1/2}/K_2$, provide very linear Arrhenius plots for

TABLE 7.3. Optimized rate parameters for 0.98% Cu/activated carbon[a]
(Reprinted from ref. 11, copyright © 2003, with permission from Elsevier)

Temperature (K)	k (μmol s^{-1}g cat^{-1})	K_{IPA} (atm^{-1})	$K' \times 10^{-12}$	K_{H_2} (atm^{-1})	K_{ACE} (atm^{-1})
443	0.73	41.2	6.27	0.103	25.8
448	1.63	37.4	3.41	0.062	18.1
458	2.92	30.5	2.12	0.047	12.9
473	6.89	21.0	0.86	0.027	6.9

[a] HTT-H$_2$ carbon sample

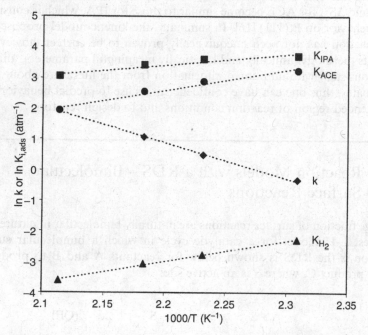

FIGURE 7.8. Rate constant, k, and adsorption equilibrium constants, K_i, from Table 7.3 versus inverse temperature. (Reprinted from ref. 11, copyright © 2003, with permission from Elsevier)

k, K_{IPA}, K_{H_2} and K_{ACE} [11], as shown in Figure 7.8, from which an activation energy of 22.9 kcal mole^{-1} can be calculated for the RDS. Table 7.4 lists the values for the standard enthalpies and entropies of adsorption, and one sees that they all conform to the criteria in Table 6.9. In addition, k_1 per site is less than 1 s^{-1}, which is far below 10^{13} s^{-1} (Table 6.10). The enthalpy for H$_2$ adsorption on Cu agrees with reported integral values but is significantly less than low-coverage values obtained for H$_2$ adsorption on Cu single crystals, whereas this enthalpy value for acetone on Cu is very similar to that reported for acetone on a Pt(111) surface, i.e., 12 kcal mole^{-1} [15]. This provides additional support for the reasonableness of these parameters.

TABLE 7.4. Thermodynamic parameters for adsorption K_i values from Eq. 9 (Reprinted from ref. 11, copyright © 2003, with permission from Elsevier)

Species	ΔH_{ad}° (kcal mole^{-1})	ΔS_{ad}° (cal mole^{-1}K^{-1})	S_g° (cal mole^{-1}K^{-1})
Isopropyl alcohol	−6.8	−8	68
H_2	−13.4	−35	36
Acetone	−13.3	−24	71

Removal of the ISO species from the site balance gives a similar fit to the data and the only significant change to the parameters in Table 7.4 is that ΔH_{ad}° and ΔS_{ad}° for ACE become similar to those for IPA, which is consistent with behavior on Pt(111) [16]. In summary, the kinetic model proposed for this reaction has not been unequivocally proven to be correct; however, it exhibits thermodynamically and kinetically meaningful parameters, and it is also consistent with additional information from the literature about these adsorbates, thus one can have confidence in its use to predict behavior over an extended region of reaction conditions and to design reactors.

7.2   Reaction Models with a RDS – Bimolecular Surface Reactions

A large fraction of surface reactions are naturally bimolecular in nature. The simplest L-H model for a catalytic cycle in which a bimolecular surface reaction is the RDS is shown below for reactants A and B to produce a single product C, where S is an active site:

$$(7.14) \qquad A + S \; \underset{}{\overset{K_A}{\rightleftharpoons}} \; A - S \qquad (QE)$$

$$(7.15) \qquad B + S \; \underset{}{\overset{K_B}{\rightleftharpoons}} \; B - S \qquad (QE)$$

$$(7.16) \qquad A - S + B - S \; \underset{\overleftarrow{k}}{\overset{\overrightarrow{k}}{\rightleftharpoons}} \; C - S + S \qquad (RDS)$$

$$(7.17) \qquad \underline{C - S \; \underset{}{\overset{K_c'}{\rightleftharpoons}} \; C + S} \qquad (QE)$$

$$A + B \; \rightleftharpoons \; C$$

Again, $K_c' = 1/K_c$, where K_c is the adsorption equilibrium constant for C. In its reversible form, again from the law of mass action for an elementary step, the net rate is:

$$r = \frac{-d[A]}{dt} = \vec{r} - \overleftarrow{r} = \vec{k}[A-S][B-S] - \overleftarrow{k}[C-S][S] \qquad (7.17)$$

Remember that $[A-S] = L\theta_A$, $[B-S] = L\theta_B$, $[C-S] = L\theta_C$ and $[S] = L\theta_v$, so the site balance is:

$$L = [S] + [A-S] + [B-S] + [C-S] \qquad (7.18)$$

and if surface concentrations are used, the probability factor for nearest-neighbor sites $(1/2\ Z/L)$ must be included and L^2 is replaced with $1/2\ ZL$ (Z is the number of nearest-neighbor sites and might be expected to range from 4 to 8). Alternatively, one can eliminate this consideration by writing the rate directly based on fractional surface coverages, i.e.,

$$r = L\vec{k}\,\theta_A\theta_B - L\overleftarrow{k}\,\theta_C\theta_v \qquad (7.19)$$

and a site balance of

$$1 = \theta_v + \theta_A + \theta_B + \theta_C \qquad (7.20)$$

is used. Consequently, utilizing a Langmuir isotherm for each species and substituting these relationships into equations 7.17–7.20, the most general form of a L-H rate expression for a bimolecular surface reaction is:

$$r = \frac{L(\vec{k}\,K_A K_B P_A P_B - \overleftarrow{k}\,K_C P_C)}{(1 + K_A P_A + K_B P_B + K_C P_C)^2} \qquad (7.21)$$

where $1/2\ Z$, which is near unity, is incorporated into the two rate constants. For low conversions (low P_C and low θ_C) and/or very small $\overleftarrow{k}$, the irreversible forward rate law becomes:

$$r = \frac{L(\vec{k}\,K_A K_B P_A P_B)}{(1 + K_A P_A + K_B P_B)^2} \qquad (7.22)$$

and an additional fitting parameter is now present for the second reactant. Over a wide enough range of partial pressure for either reactant, a rate dependence such as that shown in Figure 7.1b is expected.

The rate expression given by equation 7.21 (or 7.22) is easily modified to accommodate dissociative adsorption to allow the surface reaction of atomic species, if required. For example, if step 7.14 were written instead as:

(7.18) $$A_2 + 2S \overset{K_{A_2}}{\rightleftharpoons} 2A-S,$$

θ_A would be represented by $K_{A_2}^{1/2} P_{A_2}^{1/2}/(1 + \Sigma K_i P_i)$ with $K_i P_i = K_{A_2}^{1/2} P_{A_2}^{1/2}$ for A atoms. Keep in mind that the site balance used to obtain the denominator in the preceding sentence can contain other species, such as an inhibitor, that are not directly involved in the reaction, as discussed in Chapter 5.3.4. Using this term in the site balance (Equation 7.18 or 7.20), the reversible rate expression now becomes:

$$r = \frac{L(\vec{k}\,K_{A_2}^{1/2}K_B P_{A_2}^{1/2}P_B - \overleftarrow{k}\,K_C P_C)}{(1 + K_{A_2}^{1/2}P_{A_2}^{1/2} + K_B P_B + K_C P_C)^2} \tag{7.23}$$

As with a unimolecular reaction, it is possible to linearize either equation 7.22 or 7.23 by holding one reactant concentration constant as the other is varied; for example, the reciprocal of the rate law can be used. With the former equation, after both reactant concentrations are varied, the two plots of $(P_i/r)^{1/2}$ vs. P_i provide enough information to determine Lk, K_A and K_B.

One other possibility exists for an ideal surface in that two different types of active sites might exist, with one set of sites adsorbing only one of the reactants and the other set of sites adsorbing another reactant with or without competing adsorption from the first reactant. Thus, for example, one could have:

$$(7.19) \qquad A + S_1 \overset{K_A}{\rightleftharpoons} A - S_1 \qquad (QE)$$

$$(7.20) \qquad B + S_2 \overset{K_B}{\rightleftharpoons} B - S_2 \qquad (QE)$$

$$(7.21) \qquad A - S_1 + B - S_2 \overset{\vec{k}}{\underset{\overleftarrow{k}}{\rightleftharpoons}} C - S_2 + S_1 \qquad (RDS)$$

$$(7.22) \qquad \underline{C - S_2 \overset{1/K_C}{\rightleftharpoons} C + S_2} \qquad (QE)$$
$$\qquad\qquad A + B \rightleftharpoons C$$

For convenience, if an irreversible reaction is considered ($\vec{r} \gg \overleftarrow{r}$), then

$$r = -d[A]/dt = \vec{k}[A - S_1][B - S_2] = L\vec{k}\,\theta_A\theta_B \tag{7.24}$$

but now a separate site balance must be used for each set of sites, and because A and B do not compete for one set of sites, one has, for example

$$L_1 = [S_1] + [A - S_1] \tag{7.25}$$

and

$$L_2 = [S_2] + [B - S_2] + [C - S_2] \tag{7.26}$$

The appropriate form of the Langmuir isotherm determined from equations 7.25 and 7.26 for each set of sites is substituted into equation 7.24, which gives:

$$r = \frac{L_1 L_2 \vec{k}\,K_A K_B P_A P_B}{(1 + K_A P_A)(1 + K_B P_B + K_C P_C)} \tag{7.27}$$

The principal result here is that no pressure (or concentration) term in the denominator is squared. Also, the rate constant $\bar{k}$ will incorporate some probability factor to recognize that S_1 and S_2 must be nearest neighbor sites for step 7.21 to occur. One of the most probable situations for a reaction sequence like this pertains to hydrogenation or dehydrogenation reactions, such as those involving hydrocarbons and organic molecules, because the H atom is very small and prefers to sit in high-coordination 3-fold and 4-fold hollow sites on a metal surface, whereas the larger hydrocarbon molecule sits on one or more on-top metal sites above the H atoms [17,18]. This is discussed in greater detail in Chapter 6.3.2.

Additional simplification is acquired when assumptions are made regarding the surface species to be included in the site balance. Frequently it is neither possible nor necessary to include every intermediate because many concentrations are going to be very small compared to others. By selecting those expected to dominate based upon the behavior of the rate and any spectroscopic information, when available, the number of unknown surface coverages can be significantly decreased. At the limit, if a MARI exists, then only one surface species becomes important in the final rate derivation. In all cases with a reaction sequence involving Langmuir isotherms to relate surface coverages to partial pressures of an adsorbate, the assumption that the coverage of a particular intermediate is negligible compared to others results in the exclusion of this surface species in the site balance and the loss of this corresponding term in the denominator of the rate law. If the surface is very highly covered with one or more species and thus near saturation, the vacant site term can also be excluded. This reminder is provided to emphasize that there is a direct correlation between the form of the rate expression and the physical state of the surface, and it must be consistent.

Limiting cases of equations 7.21 and 7.22 can be expected depending on the relative sizes of θ_A, θ_B, θ_C and θ_v, and some early examples of bimolecular surface reactions have been collected from the literature by Laidler [3]. For example:

a) If A, B, and C are weakly adsorbed, then θ_A, θ_B, $\theta_C \ll \theta_v([S] \cong L)$ and $r = kP_A P_B$. This has been reported for ethylene hydrogenation on Cu between 423 and 573 K.

b) If A and C are weakly adsorbed and adsorbed B is the MARI (θ_A, $\theta_C \ll \theta_B$, θ_v) then $r = kP_A P_B/(1 + K_B P_B)^2$. An equation of this form was obtained for the reverse water gas shift reaction, $CO_2 + H_2 \rightarrow CO + H_2O$, on Pt at 1173–1373 K in which CO_2 was the MARI.

c) If A and C are again weakly adsorbed but B is now so strongly adsorbed that it nearly saturates the surface ($\theta_B \gg \theta_A$, θ_C, θ_v), then $r = kP_A/P_B$. An example of this was provided by CO oxidation on quartz between 473 and 573 K, with CO being the strongly adsorbed reactant.

A very nice, more recent example of a bimolecular surface reaction is provided by NO decomposition on an oxide surface, and the details of this system are given in Illustration 7.3.

Illustration 7.3 – NO Decomposition on Mn_2O_3: A Bimolecular L-H Reaction Model

The decomposition of nitric oxide (NO) to N_2 and O_2 on Mn_2O_3 and Mn_3O_4 has been examined in detail by Yamashita and Vannice [19], and reaction orders on NO varying from 1.4 to 1.6 between 833 and 873 K established that this was not a 1st-order reaction. Treatment of the two oxides at 873 K removed small amounts of oxygen from the lattice, but did not alter the bulk XRD patterns, and prior to reaction the surface areas were $30.6 \, m^2 \, g^{-1}$ for Mn_2O_3 and $18.3 \, m^2 \, g^{-1}$ for Mn_3O_4. After this treatment, O_2 and NO chemisorption was measured at 295 K, with the former being activated and small, whereas values for the irreversible NO uptake were $4.7 \, \mu mole \, m^{-2}$ for Mn_2O_3 and $4.1 \, \mu mole \, m^{-2}$ for Mn_3O_4. Between 793 and 873 K, an apparent activation energy of $11 \, kcal \, mole^{-1}$ was obtained for Mn_2O_3 and $15 \, kcal \, mole^{-1}$ was obtained for Mn_3O_4 from the Arrhenius plots given in Figure 7.9. Partial pressure studies with Mn_2O_3 under differential reactor conditions at different temperatures, as shown in Figures 7.10 and 7.11, gave the reaction orders listed in Table 7.5.

A L-H model with a bimolecular surface reaction between two adsorbed NO molecules as the RDS was proposed as follows, where S is an active site and the stoichiometric number for an elementary step lies outside the brackets around that step, i.e.,

$$(1) \quad 2[NO + S \underset{}{\overset{K_{NO}}{\rightleftharpoons}} NO - S]$$

$$(2) \quad 2 \, NO - S \overset{k}{\longrightarrow} N_2 + 2 \, O - S \, (RDS) \quad or \quad \left\{ \begin{array}{l} 2 \, NO - S \overset{k'}{\longrightarrow} N_2O - S + O - S \\ N_2O - S \overset{k''}{\longrightarrow} N_2 + O - S \end{array} \right\}$$

$$(3) \quad \underline{2 \, O - S \underset{}{\overset{1/K_{O_2}}{\rightleftharpoons}} O_2 + 2 \, S}$$

$$(4) \quad 2 \, NO \implies N_2 + O_2$$

The RDS shown in the cycle (step 2) is very likely more accurately represented by the irreversible 2-step sequence in the brackets to its right, but this

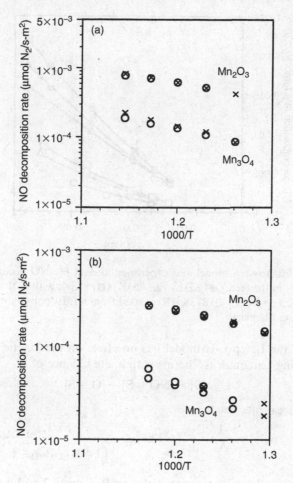

FIGURE 7.9. Arrhenius plots for NO decomposition on Mn_2O_3 and Mn_3O_4: (a) $P_{NO} = 0.04$ atm, (b) $P_{NO} = 0.02$ atm; (o) ascending temperature, (×) descending temperature. (Reprinted from ref. 19, copyright © 1996, with permission from Elsevier)

TABLE 7.5. Power law reaction orders with respect to NO
and O_2 for NO decomposition (Reprinted from ref. 19, copyright © 1996,
with permission from Elsevier)

Catalyst	Temperature (K)	NO[a]	NO[b]	O_2[c]
Mn_2O_3	833	1.4	1.6	−0.4
	853	1.5	1.9	−0.3
	873	1.5	1.8	−0.3
Mn_3O_4	873	1.6	—	—

[a] No O_2 in the feed gas.
[b] 0.01 atm O_2 in the feed gas.
[c] $P_{No} = 0.02$ atm.

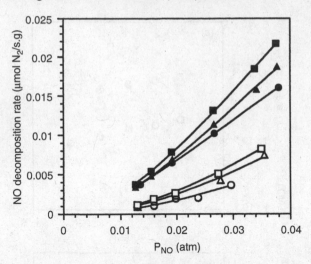

FIGURE 7.10. Fit between model and experimental data for NO decomposition on Mn_2O_3. No O_2 in the feed: (●) 833K, (▲) 853K, (■) 873K; with 0.01 atm O_2 in the feed: (○) 833K, (△) 853K, (□) 873K. (Reprinted from ref. 19, copyright © 1996, with permission from Elsevier)

alteration of the 1-step L-H model has no effect on the kinetic rate expression and, using Langmuir isotherms with a site balance of

$$L = [S] + [NO - S] + [O - S] \tag{1}$$

the derived areal rate law is:

$$r_a \left(\mu\text{mole } s^{-1} \, m^{-2} \right) = \frac{-1}{A} \frac{dN_{N_2}}{dt} = Lk\theta_{NO}^2 = \frac{k' K_{NO}^2 P_{NO}^2}{\left(1 + K_{NO}P_{NO} + K_{O_2}^{1/2} P_{O_2}^{1/2} \right)^2} \tag{2}$$

This expression allowed a reaction order on P_{NO} up to 2 and it provided for a strong inhibitory effect due to O_2, and the fits to the data are given by the solid lines in Figures 7.10 and 7.11. The fitting parameters obtained from the optimized rate equations are listed in Table 7.6. Arrhenius plots of these parameters provided the following values [19]: the activation energy for k in the RDS is $E = 46 \, \text{kcal mole}^{-1}$; for NO adsorption $\Delta H_{ad}^o = -25 \, \text{kcal mole}^{-1}$ and $\Delta S_{ad}^o = -25$ e.u.; and for O_2 adsorption

TABLE 7.6. Parameters from kinetic rate expression (Eq. 2)
(Reprinted from ref. 19, copyright © 1996,
with permission from Elsevier)

	Temperature		
	833K	853K	873K
K_{NO} (atm^{-1})	12.2	7.86	6.18
k' (μmol s^{-1}m^{-2} × 10^3)	9.37	19.0	33.4
K_{O_2} (atm^{-1})	279	113	107

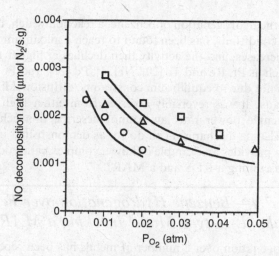

FIGURE 7.11. Fit between model and experimental data for NO decomposition on Mn_2O_3. $P_{NO} = 0.02$ atm in the feed: (○) 833K, (△) 853K, (□) 873K. (Reprinted from ref. 19, copyright © 1996, with permission from Elsevier)

$\Delta H^o_{ad} = -35$ kcal mole^{-1} and $\Delta S^o_{ad} = -31$ e.u. Thus all the requirements in Table 6.9 are fulfilled. Furthermore, if the rate constant $k' = Lk$ is converted to the correct units for k_b in Table 6.10, i.e.,

$$k_b[NO-S]^2 = k_bL^2\theta^2_{NO} = k'\theta^2_{NO} \qquad (3)$$

then

$$k_b = k'/L^2 = A_be^{-E_b/RT} \qquad (4)$$

The highest value of k' at 873 K is 0.0334 μmole s^{-1} m^{-2}, thus

$$\frac{(0.0334\,\mu\,\text{mole})}{\text{s}\cdot\text{m}^2}\left(\frac{1\,\text{m}}{10^2\,\text{cm}}\right)^2\left(6.02\times10^{17}\,\frac{\text{molecules}}{\mu\text{mole}}\right)\Bigg/\left(10^{15}\,\frac{\text{molecule}}{\text{cm}^2}\right)^2$$

$$= A_be^{\dfrac{-46000\,\text{cal/mole}}{(1.987\,\text{cal/mole}\cdot\text{K})(873\,\text{K})}}$$

and $A_b = 6.6 \times 10^{-7}$ cm^2 s^{-1} molecule^{-1}, thus satisfying Rule 3 in Table 6.10. In addition, NO adsorption on Mn_2O_3 is known to be much stronger than that of N_2O and, indeed, Q_{ad} for NO is much greater than that reported for N_2O on Mn_2O_3 (7 kcal mole^{-1} – see Illustration 7.1). Using the irreversible NO adsorption to count active sites, a TOF for N_2 formation of 3×10^{-5} s^{-1} at 773 K and $P_{NO} = 0.02$ atm can be calculated.

It was stated in Chapter 6.5 that the formulation of a consistent reaction model to derive a rate law is a valuable and useful exercise, and it provides benefits that a power rate law cannot do. A good example of this philosophy is provided by the hydrogenation of benzene, a reaction which, under a given set of reaction conditions, has been found to reach a maximum in activity as temperature increases, and the activity then declines at higher temperatures over metals such as Pt, Re and Tc [20], Ni [21], Pd [22], and Fe [23,24]. This maximum was not due to equilibrium constraints, diffusional limitations, or poisoning because it was reversibly accessed from either high or low temperatures. A single power rate law cannot describe this behavior, but a reasonable L-H-type mechanism can do so, as demonstrated in Illustration 7.4, which also provides an example of how a complex catalytic cycle can be simplified by assuming a RDS and a MARI.

Illustration 7.4 – Benzene Hydrogenation over Fe: A Bimolecular L-H-type Model Invoking a MARI

Benzene hydrogenation over a number of metals has been repeatedly investigated, and a very interesting kinetic trend had been reported in early studies showing that the catalytic activity increases normally with increasing temperature up to a point, but it then declines continuously as the temperature increases further [20,21,23]. No good explanation had been provided at the time, so a detailed kinetic study was undertaken by Yoon and Vannice to examine this reaction over Fe catalysts [24]. These Fe catalysts were characterized by XRD and CO chemisorption at 195 K [24,25], and Table 7.7 lists these irreversible CO uptakes along with the Fe crystallite sizes calculated using an adsorption stoichiometry of $CO_{ad}/Fe_s = 1/2$ and the relationship $\bar{d}_s(nm) = 0.75/D$ (see Chapter 3.3.5). These Fe catalysts exhibited significant deactivation with time on stream, so a special bracketing method was used during the kinetic runs to reactivate the catalyst and to correct for any activity loss. Differential reactor operation was employed to keep conversions below 5%, and application of the Weisz parameter gave values of 2×10^{-3} or lower [26], which verified the absence of internal diffusional effects. Very reproducible kinetic data were obtained with this approach and activity maxima vs. temperature were routinely observed near 473 K, as demonstrated by the two catalysts in Figure 7.12.

TABLE 7.7. Average Fe crystallite sizes, activities, and benzene turnover frequencies for iron catalysts (from ref. 24)

Catalyst	CO uptake (μmol/g cat)	$\bar{d}_s$ (nm)	Initial Activity[a,b] $\left(\dfrac{\mu mole\ Bz}{s \cdot g\ cat}\right)$	TOF ($s^{-1} \times 10^3$)[a] 448 K	473 K	E_{ap} (kcal mole^{-1})
4.8% Fe/GMC	27.5	12	0.44	8.0	15	23 ± 1
5.3% Fe/BC-11(2)	27.0	13	0.65	12.0	26	23 ± 2
5.2% Fe/BC-12	22.0	17	0.88	20.0	26	23 ± 1
5.8% Fe/SiO₂	10.5	37	0.23	11.0	16	22 ± 3
4.5% Fe/V3G	5.6	54	0.068	6.1	8	17 ± 2

[a] $P_{H_2} = 680$ Torr, $P_{Bz} = 50$ Torr.
[b] At 448 K.

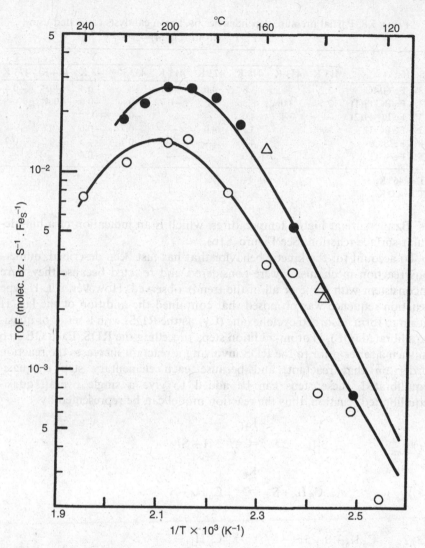

FIGURE 7.12. Turnover frequency versus 1/T (and also temperature) for iron cata-lysts; P_{H_2} = 680 Torr, P_{BZ} = 50 Torr: 4.8% Fe/GMC – (○), 5.2% Fe/BC–12: Set I – (△), Set II – (●). Solid lines represent behavior predicted using values in Table 7.9. (Reprinted from ref. 24, copyright © 1983, with permission from Elsevier)

Some of the kinetic results are also listed in Table 7.7 for two graphitized carbon supports (GMC and V3G), two boron-doped carbon supports (BC-11, BC-12), and a silica support. Partial pressure studies revealed that the dependency on benzene (Bz) was typically between zero and inverse 1st order; but the dependency on H_2 was surprisingly high and quite tempera-ture dependent, ranging above 3rd order at times, as shown in Table 7.8 and Figures 7.13 and 7.14. The latter figure also indicates an activity maximum

TABLE 7.8. Partial pressure dependencies over iron catalysts estimated using $r = kP_{H_2}^X P_B^Y$ (from ref. 24)

Catalyst	X				Y				
	413 K	433 K	448 K	473 K	413 K	433 K	448 K	474 K	493 K
4.8% Fe/GMC	3.2	—	3.8	4.0	−0.7	—	−0.4	−0.3	−0.4
5.3% Fe/BC-11(1)	2.5	3.0	3.3	3.3	−0.7	−0.6	−0.6	−0.4[a]	—
5.3% Fe/BC-11(2)	—	3.4	4.0	—	—	−0.8	−0.7[b]	—	—
5.2% Fe/BC-12	—	2.7	3.1	4.0	—	−0.7	−0.6	−0.6	—
5.8% Fe/SiO$_2$	—	—	4.0	—	—	−0.7	—	−0.2	−0.2
4.5% Fe/V3G	—	—	3.0	—	—	—	−0.4	—	—

[a] $T = 463$ K.
[b] $T = 453$ K.

vs. Bz pressure at higher temperatures, which is an indication of a bimolecular surface reaction (See Figure 7.1b).

To account for the kinetic behavior that has just been described, numerous reaction mechanisms were considered and rejected because they were inconsistent with some or all of the trends observed. However, a L-H-type reaction sequence was proposed that contained the addition of the last H atom to form adsorbed cyclohexane (Cy) as the RDS, with a series of quasi-equilibrated (QE), H atom-addition steps preceding the RDS. Each QE step in such a series prior to the RDS involving a reactant increases the reaction order on that reactant, and because each elementary step is quasi-equilibrated, these steps can be added to give a single overall quasi-equilibrated reaction, thus the reaction model can be represented by:

$$(1) \qquad 3[H_2 + 2S \underset{}{\overset{K_{H_2}}{\rightleftharpoons}} 2H-S]$$

$$(2) \qquad C_6H_6 + S \underset{}{\overset{K_{Bz}}{\rightleftharpoons}} C_6H_6-S$$

$$(3) \qquad C_6H_6 + S + 5H-S \underset{}{\overset{K_E}{\rightleftharpoons}} C_6H_{11}-S + 5S$$

$$(4) \qquad C_6H_{11}-S + H-S \overset{\vec{k}}{\longrightarrow} C_6H_{12}-S + S \qquad (RDS)$$

$$(5) \qquad C_6H_{12}-S \underset{}{\overset{1/K_{CY}}{\rightleftharpoons}} C_6H_{12} + S$$

$$(6) \qquad C_6H_6 + 3H_2 \Longrightarrow C_6H_{12}$$

where S is an active site, K_{H_2}, K_{Bz} and K_{Cy} are adsorption equilibrium constants, and K_E is the lumped equilibrium constant for step 3. The rate can be defined as:

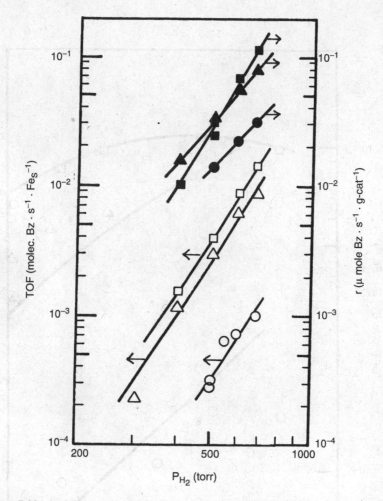

FIGURE 7.13. Activity versus hydrogen pressure for iron catalysts, $P_{BZ} = 50$ Torr: 4.8%/Fe/GMC: 413K – (○), 448K – (△), 473K – (□); 5.2% Fe/BC-12: 433K – (●), 448K – (▲), 473K – (■). (Reprinted from ref. 24, copyright © 1983, with permission from Elsevier)

$$r_m = \frac{1}{m}\frac{dN_{Cy}}{dt} = k[C_6H_{11} - S][H - S] \qquad (1)$$

Adsorption steps *1* and *2* are quasi-equilibrated, so

$$K_{H_2} = \frac{[H - S]^2}{P_{H_2}[S]^2} \qquad (2)$$

and

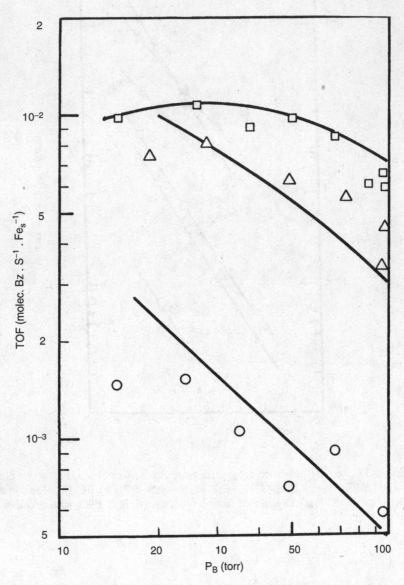

FIGURE 7.14. Activity versus benzene pressure for 4.8% Fe/GMC, $P_{H_2} = 600$ Torr: 414K – (O), 448K – ($\triangle$) 473K – ($\square$). Solid lines represent predicted behavior using parameter values in Table 7.9. (Reprinted from ref. 24, copyright © 1983, with permission from Elsevier)

$$K_{Bz} = \frac{[C_6H_6 - S]}{P_{Bz}[S]} \qquad (3)$$

Reaction 3 is also quasi-equilibrated, so

$$K_E = \frac{[C_6H_{11} - S][S]^5}{[C_6H_6 - S][H - S]^5} \qquad (4)$$

Solving equations 2 and 3 for [H – S] and [C₆H₆ – S], respectively, substituting these values into equation 4, and rearranging it gives

$$[C_6H_{11} - S] = K_{H_2}^{5/2} K_{Bz} K_E P_{H_2}^{5/2} P_{Bz}[S] \qquad (5)$$

It is now assumed that adsorbed benzene is the MARI, which is consistent with other information [24], thus

$$L = [S] + [C_6H_6 - S] \qquad (6)$$

and substituting this in equation 3 gives

$$[S] = L/(1 + K_{Bz}P_{Bz}) \qquad (7)$$

Finally, substitution of equations 2, 5 and 7 into equation 1 yields the following rate law, after including the probability of nearest neighbor sites and dividing r_m by the Fe_s concentration (per g) from Table 7.7 to convert the rate to TOF units:

$$r(s^{-1}) = \frac{L\bar{k}K_{H_2}^3 \, K_{Bz} \, K_E \, P_{H_2}^3 \, P_{Bz}}{(1 + K_{Bz}P_{Bz})^2} = \frac{k_{ap} \, P_{H_2}^3 \, P_{Bz}}{(1 + K_{Bz}P_{Bz})^2} \qquad (8)$$

The apparent activation energy associated with the apparent rate constant, k_{ap}, is the sum of four contributions, i.e.,

$$E_{ap} = E_{RDS} + 3\Delta H_{H_2}^\circ + \Delta H_{Bz}^\circ + \Delta H_E \qquad (9)$$

where E_{RDS} is the activation energy of the RDS, $\Delta H_{H_2}^\circ$ and ΔH_{Bz}° are the respective enthalpies of adsorption for H_2 and Bz, and ΔH_E is the lumped enthalpy change associated with step 3. Consequently, equation 8 can be written as

$$r = \frac{A_{ap}e^{-E_{ap}/RT}P_{H_2}^3 P_{Bz}}{\left[1 + e^{\Delta S_{Bz}^\circ/R}e^{-\Delta H_{Bz}^\circ/RT} \cdot P_{Bz}\right]^2} \qquad (10)$$

This equation predicts a maximum in the rate as the temperature varies, as shown by taking the derivative of this expression [24]. One can understand this qualitatively because $\Delta H_{H_2}^\circ$ and ΔH_{Bz}° are negative, thus as the temperature increases the surface coverages of H atoms and benzene molecules drop significantly, and at some point the latter declines so much that the Bz surface coverage term decreases more rapidly than the rate constant k increases, and the apparent "activation energy" becomes negative when

TABLE 7.9. Optimized kinetic parameter values for iron catalysts (from ref. 24)

Catalyst	A (Bz molec.s^{-1} Fe$_s^{-1}$atm^{-4})	E_{ap} (kcal mole^{-1})	ΔH_{Bz}^o (kcal mole^{-1})	ΔS_{Bz}^o (e.u.)	E'_{ap} (kcal mole^{-1})[a] Predicted	Obs.
4.8% Fe/GMC	3.02×10^{-13}	−28	−24	−44	20	23
5.3% Fe/BC-11(2)	5.23×10^{-13}	−28	−24	−44	20	23
5.2% Fe/BC-12	3.17×10^{-13}	−28	−25	−47	22	23
5.8% Fe/SiO$_2$	6.86×10^{-12}	−25	−22	−40	19	22
4.5% Fe/V3G	3.78×10^{-9}	−18	−17	−30	16	17

[a] E'_{ap} is the apparent activation energy in the low-temperature region.

the absolute value of $3\Delta H_{H_2}^o + \Delta H_{Bz}^o$ is greater than that of $E_{RDS} + \Delta H_E$. The four parameters in equation 10 were optimized and the results for the five catalysts in Table 7.7 are listed in Table 7.9. At lower temperatures, the surface will be nearly saturated with Bz and $K_{Bz}P_{Bz} \gg 1$; in this regime the apparent activation energy, E'_{ap}, is approximately $E'_{ap} \cong E_{ap} - 2\Delta H_{Bz}^o$. These predicted E'_{ap} values are compared to those observed in the two right-hand columns in Table 7.9, and the agreement is consistently close. Finally, the ΔH_{Bz}^o and ΔS_{Bz}^o values satisfy all the criteria in Table 6.9. It is true that reaction orders in H_2 that are greater than three cannot be explained by this model; however, subsequent studies of Pd and Pt catalysts have shown that reaction orders on H_2 above 3 can be obtained if concurrent dehydrogenation reactions occur to create carbonaceous species on the surface which inhibit this hydrogenation reaction [22,27].

7.3   Reaction Models with a RDS – Reactions between an Adsorbed Species and a Gas-Phase Species

This type of reaction sequence requires that only one of the two reactants be adsorbed on the surface. It was also considered by Langmuir but was not readily accepted until revived by Rideal [28], thus although appropriately referred to as the Langmuir-Rideal mechanism in some texts [29], it is more commonly known as the Rideal-Eley (R-E) mechanism [30]. This mechanism assumes that the making and breaking of bonds occurs during the lifetime of one collision at the surface, thus it might be anticipated that its applicability would be restricted to extremely reactive gas-phase species, such as free radicals, ions and certain atoms. A more likely scenario is that in reactions that appear to follow R-E kinetic behavior, the gas-phase species adsorbs in a weakly-bound precursor state to allow sufficient residence time on the surface for the appropriate bonds to be formed and energy transfer to occur. These low coverages would also give a 1st-order dependence on the gas-phase reactant, i.e., it would constitute coverages in the Henry's law region of the Langmuir isotherm.

This author is unaware of any unequivocal, proven R-E reaction mechanisms in the literature and is therefore hesitant to use it; however, this mechanism is routinely referred to in texts on kinetics and for completeness it is represented below:

$$\text{(7.23)} \qquad A + S \; \overset{K_A}{\rightleftharpoons} \; A - S \qquad \text{(QE)}$$

$$\text{(7.24)} \qquad B + A - S \; \overset{\vec{k}}{\longrightarrow} \; C - S \qquad \text{(RDS)}$$

$$\text{(7.25)} \qquad \underline{C - S \; \overset{1/K_C}{\rightleftharpoons} \; C + S} \qquad \text{(QE)}$$

$$A + B \; \rightleftharpoons \; C$$

Here, if step 7.24 is considered to be irreversible:

$$r = -\frac{d[A]}{dt} = \vec{k} \, P_B [A - S] = L \vec{k} \, P_B \theta_A \tag{7.28}$$

and with only A and C in the site balance, i.e., $1 = \theta_v + \theta_A + \theta_C$, the final rate expression is:

$$r = L \vec{k} \, K_A P_A P_B / (1 + K_A P_A + K_C P_C) \tag{7.29}$$

The denominator is not squared, there is always a 1st-order dependence on B, and the dependence on A is similar to that of a unimolecular reaction, as shown in Figure 7.1c. Sometimes B has been included in the site balance, but, if so, adsorbed B is assumed to be inactive.

Finally, to summarize, a variety of the traditional rate expressions for reactions on an ideal surface has been examined, and many of their derivations have been discussed in detail. These include L-H and R-E models describing unimolecular and bimolecular reactions on surfaces with either one type of active site or two types of active sites. If a RDS other than that for a surface reaction is proposed, i.e., either an adsorption or a desorption step, then a H-W rate expression is derived. These standard rate laws, which assume a RDS exists, are frequently referred to and utilized, and they are summarized in Table 7.10. Many other forms of a rate expression, which do not assume a RDS and utilize the SSA, can be derived based on the reaction sequence proposed.

7.4   Reaction Models with no RDS

As mentioned at the beginning of this chapter, there are reactions which, under certain conditions, may not have a RDS; if so, the sequence of elementary steps in the catalytic cycle must reflect this. These sequences may be comprised of two or more slow steps along with one or more

TABLE 7.10. Langmuir-Hinshelwood (L–H), Hougen-Watson (H-W) Gas-Phase Rate Expressions: L is the site density, S is an active site, K_i is either the equilibrium constant for step i or the adsorption equilibrium constant for species i, and K is the equilibrium constant for the overall reaction

General reaction model	Rate determining step	Rate expression
Unimolecular – Single Site		
(1) $A + S \underset{k_{-1}}{\overset{k_1}{\rightleftharpoons}} A \cdot S$	Adsorption Step 1	$r = \dfrac{Lk_1P_A - Lk_{-1}P_B/K_2K_3}{(1+P_B/K_2K_3+P_B/K_3)} = \dfrac{Lk_1(P_A - P_B/K)}{(1+K_AP_B/K+K_BP_B)}$
(2) $A \cdot S \underset{k_{-2}}{\overset{k_2}{\rightleftharpoons}} B \cdot S$	Surface Reaction Step 2	$r = \dfrac{Lk_2K_1P_A - Lk_{-2}P_B/K_3}{(1+K_1P_A+P_B/K_3)} = \dfrac{Lk_2K_A(P_A - P_B/K)}{(1+K_AP_A+K_BP_B)}$
(3) $B \cdot S \underset{k_{-3}}{\overset{k_3}{\rightleftharpoons}} B + S$	Desorption Step 3	$r = \dfrac{Lk_3K_1K_2P_A - Lk_{-3}P_B}{(1+K_1P_A+K_1K_2P_A)} = \dfrac{Lk_{-3}(KP_A - P_B)}{(1+K_AP_A+K_BKP_A)}$
$A \rightleftharpoons B$		
Unimolecular – Dual Site		
(1) $A + S \underset{k_{-1}}{\overset{k_1}{\rightleftharpoons}} A \cdot S$	Adsorption Step 1	$r = \dfrac{Lk_1P_A - Lk_{-1}P_BP_C/K_2K_3K_4}{(1+P_BP_C/K_2K_3K_4+P_B/K_3+P_C/K_4)} = \dfrac{Lk_1(P_A - P_BP_C/K)}{(1+K_AP_BP_C/K+K_BP_B+K_CP_C)}$
(2) $A \cdot S + S \underset{k_{-2}}{\overset{k_2}{\rightleftharpoons}} B \cdot S + C \cdot S$	Surface Reaction Step 2	$r = \dfrac{Lk_2K_1P_A - Lk_{-2}P_BP_C/K_3K_4}{(1+K_1P_A+P_B/K_3+P_C/K_4)^2} = \dfrac{Lk_2K_A(P_A - P_BP_C/K)}{(1+K_AP_A+K_BP_B+K_CP_C)^2}$
(3) $B \cdot S \underset{k_{-3}}{\overset{k_3}{\rightleftharpoons}} B + S$ (4) $C \cdot S \underset{k_{-4}}{\overset{k_4}{\rightleftharpoons}} C + S$	Desorption Step 4 (or 3)	$r = \dfrac{Lk_4K_1K_2K_3P_A/P_B - Lk_{-4}P_C}{(1+K_1P_A+P_B/K_3+K_1K_2K_3P_A/P_B)} = \dfrac{Lk_{-4}(KP_A/P_B - P_C)}{(1+K_AP_A+K_BP_B+K_CKP_A/P_B)}$
$A \rightleftharpoons B + C$		

Bimolecular-Nondissociative

(1) $A + S \underset{k_{-1}}{\overset{k_1}{\rightleftharpoons}} A \cdot S$ Adsorption Step 1 (or 2)

$$r = \frac{Lk_1 P_A - Lk_{-1} P_C P_D / K_2 K_3 K_4 K_5 P_B}{\left(1 + P_C P_D / K_2 K_3 K_4 K_5 P_B + K_2 P_B + P_C / K_4 + P_D / K_5\right)} = \frac{Lk_1 (P_A - P_C P_D / K P_B)}{\left(1 + \frac{K_A P_C P_D}{K P_B} + K_B P_B + K_C P_C + K_D P_D\right)}$$

(2) $B + S \underset{k_{-2}}{\overset{k_2}{\rightleftharpoons}} B \cdot S$

(3) $A \cdot S + B \cdot S \underset{k_{-3}}{\overset{k_3}{\rightleftharpoons}} C \cdot S + D \cdot S$ Surface Reaction Step 3

$$r = \frac{Lk_3 K_1 K_2 P_A P_B - Lk_{-3} P_C P_D / K_4 K_5}{\left(1 + K_1 P_A + K_2 P_B + P_C / K_4 + P_D / K_5\right)^2} = \frac{Lk_3 K_A K_B (P_A P_B - P_C P_D / K)}{\left(1 + K_A P_A + K_B P_B + K_C P_C + K_D P_D\right)^2}$$

(4) $C \cdot S \underset{k_{-4}}{\overset{k_4}{\rightleftharpoons}} C + S$ Desorption Step 5 (or 4)

(5) $D \cdot S \underset{k_{-5}}{\overset{k_5}{\rightleftharpoons}} D + S$

$$r = \frac{Lk_5 K_1 K_2 K_3 K_4 P_A P_B / P_C - Lk_{-5} P_D}{\left(1 + K_1 P_A + K_2 P_B + P_C / K_4 + K_1 K_2 K_3 K_4 P_A P_B / P_C\right)} = \frac{Lk_{-5} (K P_A P_B / P_C - P_D)}{\left(1 + K_A P_A + K_B P_B + K_C P_C + K_D K P_A P_B / P_C\right)}$$

$\overline{A + B \rightleftharpoons C + D}$

Bimolecular-Dissociative

(1) $A_2 + 2S \underset{k_{-1}}{\overset{k_1}{\rightleftharpoons}} 2A \cdot S$ Adsorption Step 1

$$r = \frac{Lk_1 P_{A_2} - Lk_{-1} (P_C / K_2 K_3 K_4 P_B)^2}{\left(1 + P_C / K_2 K_3 K_4 P_B + K_2 P_B + P_C / K_4\right)^2} = \frac{Lk_1 \left(P_{A_2} - P_C^2 / K P_B^2\right)}{\left(1 + K_A^{1/2} P_C / K^{1/2} P_B + K_B P_B + K_C P_C\right)^2}$$

(2) $2[B + S \underset{k_{-2}}{\overset{k_2}{\rightleftharpoons}} B \cdot S]$

(3) $2[A \cdot S + B \cdot S \underset{k_{-3}}{\overset{k_3}{\rightleftharpoons}} C \cdot S + S]$ Surface Reaction Step 3

$$r = \frac{Lk_3 K_1^{1/2} K_2 P_{A_2}^{1/2} P_B - Lk_{-3} P_C / K_4}{\left(1 + K_1^{1/2} P_{A_2}^{1/2} + K_2 P_B + P_C / K_4\right)^2} = \frac{Lk_3 K_{A_2}^{1/2} K_B \left(P_{A_2}^{1/2} P_B - P_C / K^{1/2}\right)}{\left(1 + K_{A_2}^{1/2} P_{A_2}^{1/2} + K_B P_B + K_C P_C\right)^2}$$

(4) $2[C \cdot S \underset{k_{-4}}{\overset{k_4}{\rightleftharpoons}} C + S]$ Desorption Step 4

$$r = \frac{Lk_4 K_1^{1/2} K_2 K_3 P_{A_2}^{1/2} P_B - Lk_{-4} P_C}{\left(1 + K_1^{1/2} P_{A_2}^{1/2} + K_2 P_B + K_1^{1/2} K_2 K_3 P_{A_2}^{1/2} P_B\right)} = \frac{Lk_{-4} \left(K_{A_2}^{1/2} P_{A_2}^{1/2} P_B - P_C\right)}{\left(1 + K_{A_2}^{1/2} P_{A_2}^{1/2} + K_B P_B + K_C K P_{A_2}^{1/2} P_B\right)}$$

$\overline{A_2 + 2B \rightleftharpoons 2C}$

Continued

TABLE 7.10. (continued)

General reaction model	Rate determining step	Rate expression
Bimolecular-Different Sites		
(1) $A + S_1 \underset{k_1}{\overset{k_1}{\rightleftharpoons}} A \cdot S_1$	Adsorption Step 1	$r = \dfrac{L_1 k_1 P_A - L_1 k_{-1} P_C / K_2 K_3 K_4 P_B}{(1 + P_C / K_2 K_3 K_4 P_B + P_C / K_4)} = \dfrac{L_1 k_1 (P_A - P_C / K P_B)}{(1 + K_A P_C / K P_B + K_C P_C)}$
(2) $B + S_2 \underset{k_2}{\overset{k_2}{\rightleftharpoons}} B \cdot S_2$	Adsorption Step 2	$r = \dfrac{L_2 k_2 P_B - L_2 k_{-2} P_C / K_1 K_3 K_4 P_A}{(1 + P_C / K_1 K_3 K_4 P_A)} = \dfrac{L_2 k_2 (P_B - P_C / K P_A)}{(1 + K_B P_C / K)}$
(3) $A \cdot S_1 + B \cdot S_2 \underset{k_{-3}}{\overset{k_3}{\rightleftharpoons}} C \cdot S_1 + S_2$	Surface Reaction Step 3	$r = \dfrac{L_1 L_2 k_3 K_1 K_2 P_A P_B - L_1 L_2 k_{-3} P_C / K_4}{(1 + K_1 P_A + P_C / K_4)(1 + K_2 P_B)} = \dfrac{L' k_3 K_1 K_2 (P_A P_B - P_C / K)}{(1 + K_A P_A + K_C P_C)(1 + K_B P_B)}$
(4) $C \cdot S_1 \underset{k_{-4}}{\overset{k_4}{\rightleftharpoons}} C + S_1$	Desorption Step 4	$r = \dfrac{L_1 k_4 K_1 K_2 K_3 P_A P_B - L_1 k_{-4} P_C}{(1 + K_1 P_A + K_1 K_2 K_3 P_A P_B)} = \dfrac{L_1 k_{-4} (K P_A P_B - P_C)}{(1 + K_A P_A + K_C K P_A P_B)}$
$\overline{A + B \rightleftharpoons C}$		
Bimolecular-Rideal-Eley		
(1) $A + S \underset{k_{-1}}{\overset{k_1}{\rightleftharpoons}} A \cdot S$	Adsorption Step 1	$r = \dfrac{L k_1 P_A - L k_{-1} P_C / K_2 K_3 P_B}{(1 + P_C / K_2 K_3 P_B + P_C / K_3)} = \dfrac{L k_1 (P_A - P_C / K P_B)}{(1 + K_A P_C / K P_B + K_C P_C)}$
(2) $A \cdot S + B \underset{k_{-2}}{\overset{k_2}{\rightleftharpoons}} C \cdot S$	Surface Reaction Step 2	$r = \dfrac{L k_2 K_1 P_A P_B - L k_{-2} P_C / K_3}{(1 + K_1 P_A + P_C / K_3)} = \dfrac{L k_2 K_A (P_A P_B - P_C / K)}{(1 + K_A P_A + K_C P_C)}$
(3) $C \cdot S \underset{k_{-3}}{\overset{k_3}{\rightleftharpoons}} C + S$	Desorption Step 3	$r = \dfrac{L k_3 K_1 K_2 P_A / P_B - L k_{-3} P_C}{(1 + K_1 P_A + K_1 K_2 P_A / P_B)} = \dfrac{L k_{-3} (K P_A / P_B - P_C)}{(1 + K_A P_A + K_C K P_A / P_B)}$
$\overline{A + B \rightleftharpoons C}$		

quasi-equilibrated steps. In either situation, the SSA must be applied to eliminate the unknown concentrations of surface intermediates contained in the site balance, which must still be used to eliminate all of the unknown species that are considered. Some of the parameters in the denominator of the rate equation will now represent ratios of rate constants or groups of rate constants. It must not be forgotten, however, that if the concentration of an intermediate is governed by more than one elementary step, but one of these steps is quasi-equilibrated, this latter step will dominate and control the concentration. In many cases, the rate equation that results from this derivation can have a mathematical form that is identical to that obtained from a L-H rate law [31].

7.4.1   A Series of Irreversible Steps – General Approach

One relatively common sequence of elementary steps is that consisting of two or more irreversible steps. One example of a more complicated unimolecular reaction demonstrating this behavior is provided in Illustration 7.5.

Illustration 7.5 – Dehydrogenation of Methylcyclohexane on Pt: no RDS

One of the easiest catalytic reforming reactions – the dehydrogenation of methylcyclohexane (MCH) to toluene (TOL):

$$C_6H_{11}CH_3 \Longrightarrow C_6H_5CH_3 + 3H_2$$

was used by Sinfelt and coworkers to gain insight into reforming catalysts [32]. This reaction on a 0.3% Pt/Al_2O_3 catalyst, where the selectivity to toluene is very high (little cracking to give CH_4), was studied over the following range of reaction conditions: $T = 588–645\,K$, $P_{MCH} = 0.07–2.2\,atm$, and $P_{H_2} = 1.1–4.1\,atm$. (Question: Why were these high H_2 pressures used for a dehydrogenation reaction? Answer: To minimize deactivation by inhibiting the build-up of carbonaceous deposits on the catalyst surface.) The experimental data were fit well by the rate expression:

$$r = aP_{MCH}/(1 + bP_{MCH}) \qquad\qquad (1)$$

and the following sequence was proposed [32]:

(1) $\qquad\qquad$ $(MCH)_g \rightleftharpoons (MCH)_{ad}$

(2) $\qquad\qquad$ $(MCH)_{ad} \longrightarrow (TOL)_{ad}$ $\qquad$ (RDS)

(3) $\qquad\qquad$ $(TOL)_{ad} \rightleftharpoons (TOL)_g$

Note that this method of writing reactions without incorporating the active site, which has been frequently used in the past, can be ambiguous and create difficulties. To conform to the approach adopted in this book, this sequence should be written invoking an active site, *, as:

$$(4) \qquad MCH_g + * \overset{K_{MCH}}{\rightleftharpoons} MCH*$$

$$(5) \qquad MCH* \overset{\bar{k}}{\xrightarrow{A}} TOL* + 3H_2 \qquad (RDS)$$

$$(6) \qquad TOL* \overset{1/K_{TOL}}{\rightleftharpoons} TOL_g + *$$

$$\overline{\qquad MCH_g \Longrightarrow TOL_g + 3H_2 \qquad}$$

Now, the probability of losing 6 H atoms and forming 3 H_2 molecules in a single elementary step is essentially zero, thus one already can see that a modification is almost certainly needed. We will address this complication a little later. Ignoring this concern for now, if the sequence of steps 4–6 is used along with the assumption that MCH is the MARI, then a $= L\bar{k}\,K_{MCH}$ and b $= K_{MCH}$, and Arrhenius plots of these two parameters give values of E $= 33$ kcal mole^{-1} for $\bar{k}$, while $\Delta H^\circ_{ad} = -30$ kcal mole^{-1} and $\Delta S^\circ_{ad} = -45$ e.u. for MCH, thus $E_{ap} = E + \Delta H = 3$ kcal mole^{-1}. Both of these latter values fulfill the criteria in Table 6.9; however, this enthalpy of adsorption is too high for a cyclic paraffin on a metal surface. In addition, heats of adsorption for aromatics are known to be higher than those for paraffins, yet when benzene or m-xylene was added to the feed, rates decreased by only 20% or less, indicating little competitive adsorption was occurring. These inconsistencies with the model after this additional testing consequently led to its rejection.

Commensurate with this additional information, the assumption of a RDS was discarded and the following sequence of irreversible steps was proposed [32]:

$$(7) \qquad (MCH)_g \overset{k_1}{\longrightarrow} (MCH)_{ad}$$

$$(8) \qquad (MCH)_{ad} \overset{k_2}{\longrightarrow} (TOL)_{ad} + 3H_2$$

$$(9) \qquad (TOL)_{ad} \overset{k_3}{\longrightarrow} (TOL)_g$$

Again, to more accurately represent this sequence as a series of elementary steps, one possibility would be:

(10) $\qquad$ $C_6H_{11}CH_{3_g} + * \xrightarrow{\quad k_1 \quad} C_6H_{11}CH_3*$

(11) $\qquad$ $CH_6H_{11}CH_3* \xrightarrow{\quad k_2 \quad} C_6H_9CH_3* + H_2$

(12) $\qquad$ $C_6H_9CH_3* \xrightarrow{\quad k_2' \quad} C_6H_7CH_3* + H_2$

(13) $\qquad$ $C_6H_7CH_3* \xrightarrow{\quad k_2'' \quad} C_6H_5CH_3* + H_2$

(14) $\qquad$ $C_6H_5CH_3* \xrightarrow{\quad k_3 \quad} C_6H_5CH_{3_g} + *$

With either sequence, if adsorbed TOL is proposed as the MARI, then the site balance is:

$$1 = \theta_v + \theta_{TOL} \qquad (1)$$

Application of the SSA to adsorbed toluene gives:

$$\frac{d\lfloor C_6H_5CH_3* \rfloor}{dt} = \frac{Ld\theta_{TOL}}{dt} = k_2''[C_6H_7CH_3*] - k_3[C_6H_5CH_3*] = 0 \qquad (2)$$

Continuous application of the SSA to $C_6H_7CH_3*$ and the other reaction intermediates gives finally:

$$Lk_1P_{MCH}\theta_v - Lk_3\theta_{TOL} = 0 \qquad (3)$$

therefore,

$$\theta_{TOL} = k_1P_{MCH}\theta_v/k_3 \qquad (4)$$

and

$$\theta_v = 1/(1 + (k_1/k_3)P_{MCH}) \qquad (5)$$

Consequently, the rate (per mass of catalyst) is:

$$r_m = \frac{-dN_{MCH}}{m\,dt} = Lk_1P_{MCH}\theta_v = Lk_1P_{MCH}/(1 + (k_1/k_3)P_{MCH}) \qquad (6)$$

The apparent activation energy of 3 kcal mole^{-1} for k_1 is now reasonable for an adsorption step and $\ln b = \ln(k_1/k_3) = -(E_1 - E_3)/RT + \ln A_1/A_3$, thus the desorption energy for toluene is 33 kcal mole^{-1}, which is much more reasonable relative to its high Q_{ad}. Also, $LA_3 \ll 10^{28}$ cm^{-2}s^{-1}, which satisfies the criterion in Table 7.10. The moral of this example is: even though all available criteria may be satisfied, there is still no guarantee that the model is correct!

It was stated earlier that a sequence of elementary steps comprising a catalytic cycle need not contain a RDS, and Illustration 7.5 provides one example of this. It was also mentioned previously in this chapter that it is plausible for some catalytic systems to have two types of active sites, and catalysts used for hydrogenation reactions offer some of the best possibilities for this to occur. A good example of both of these situations appears to be that of acetic acid reduction by H_2 over Pt and Fe catalysts [33–36], as described in Illustration 7.6.

Illustration 7.6 – Acetic Acid Reduction by H_2 on Pt: The Presence of Two Types of Sites and Absence of a RDS

This reaction can be quite complex and can result in the formation of numerous products including acetaldehyde, ethanol, ethyl acetate, ethane, CH_4 and CO. As a result, this network of reactions provided an excellent opportunity to examine the role a support can play, because with Pt/SiO_2 catalysts at low conversions only hydrogenolysis occurred to form CH_4 and CO, whereas with Pt/TiO_2 catalysts ethanol (50%), ethyl acetate (30%) and ethane (20%) were produced [33]. In situ characterization using IR spectroscopy (DRIFTS) under reaction conditions combined with TPD and TPR (Temperature Programmed Desorption and Reduction, respectively) led to the identification of acyl and acetate species on the catalyst [34]; however, only the acyl species was a reactive intermediate at lower temperatures because the acetate species was too stable. This valuable information was incorporated into the reaction model.

Pt/TiO_2 is a catalyst system that exhibits MSI (Metal-Support Interaction) behavior (see Chapter 2.13). This is evidenced by a 0.69% Pt/TiO_2 catalyst whose total H_2 chemisorption declined from 21 μmole H_2 g^{-1} after a low temperature reduction (LTR) at 473 K to 3.5 μmole H_2 g^{-1} after a high-temperature reduction (HTR) at 773 K. This was not due to Pt sintering. The former uptake gives a H_{ad}/Pt_T ratio of 1.2, indicating a Pt dispersion of unity, and the irreversible H_2 uptake of 10.6 μmole g^{-1} after the LTR step still gives a high dispersion of 0.60. The uptake on a 2.01% Pt/TiO_2 (LTR) catalyst was 51 μmole H_2 g^{-1} which yielded a dispersion of 0.99. Vapor-phase kinetics were investigated under differential reactor conditions between 422–573 K, 0.13–0.92 atm H_2, and 0.0092–0.068 atm CH_3COOH. Application of the Weisz criterion gave values from 0.018 to 0.035, thus assuring the absence of diffusional limitations [37]. All Pt/TiO_2 catalysts had an apparent activation energy of 13 ± 1 kcal mole^{-1} for the addition of hydrogen to acetic acid [33].

Despite the lower chemisorption capacity after a HTR pretreatment, this catalyst was 45% more active than after a LTR pretreatment and had a TOF that was 7 times higher. These higher activities and TOFs in reactions involving the hydrogenation of C-O bonds has been attributed to the creation of new sites at the metal-titania interface as a consequence of the removal of surface oxygen atoms from the TiO_2 lattice [38,39]. After assimilating all the spectroscopic and kinetic data, the latter of which are shown in

Figure 7.15, the following reaction mechanism was proposed without a RDS and with the invocation of two types of active sites, i.e., one set (*) on the Pt surface and another set (S) on the TiO$_2$ surface. This series of elementary steps describes the initial hydrogenation reaction to form acetaldehyde via

an acyl $\left[\begin{smallmatrix} & O \\ & \| \\ R & -C- \end{smallmatrix} \right]$ intermediate as well as the subsequent hydrogenation reactions to produce ethanol and ethane. The ketonization reaction, which can also occur on the TiO$_2$ surface, is not a hydrogenation reaction. The catalytic sequence is:

$$(1) \quad H_{2(g)} + 2\,* \;\overset{K_{H_2}}{\rightleftharpoons}\; 2\,H*$$

$$(2) \quad CH_3COOH_{(g)} + 2\,* \;\overset{K_{A_c}}{\rightleftharpoons}\; CH_3COO* + H*$$

$$(3) \quad CH_3COOH_{(g)} + S\,* \;\overset{K_A}{\rightleftharpoons}\; CH_3COOH\text{-}S$$

$$(4) \quad CH_3COOH\text{-}S \;\overset{K_{Acy}}{\rightleftharpoons}\; CH_3CO\text{-}S\text{-}OH$$

$$(5) \quad CH_3CO\text{-}S\text{-}OH + H* \;\xrightarrow{k_1}\; CH_3CHO\text{-}S\text{-}OH + *$$

$$(6) \quad CH_3CHO\text{-}S\text{-}OH + H* \;\xrightarrow{k_2}\; CH_3CHO\text{-}S + H_2O_{(g)} + *$$

$$(7) \quad CH_3CHO\text{-}S \;\underset{k_{-3}}{\overset{k_3}{\rightleftharpoons}}\; CH_3CHO_{(g)} + S$$

$$(8) \quad CH_3CHO\text{-}S + H* \;\xrightarrow{k_4}\; CH_3CH_2O\text{-}S + *$$

$$(9) \quad CH_3CH_2O\text{-}S + H* \;\xrightarrow{k_5}\; CH_3CH_2OH\text{-}S + *$$

$$(10) \quad CH_3CH_2OH\text{-}S \;\underset{k_{-6}}{\overset{k_6}{\rightleftharpoons}}\; CH_3CH_2OH_{(g)} + S$$

$$(11) \quad CH_3CH_2OH\text{-}S + H* \;\xrightarrow{k_7}\; CH_3CH_2OH_2\text{-}S + *$$

$$(12) \quad CH_3CH_2OH_2\text{-}S + H* \;\xrightarrow{k_8}\; CH_3CH_3\text{-}S + H_2O_{(g)} + *$$

$$(13) \quad CH_3CH_3\text{-}S \;\underset{k_{-9}}{\overset{k_9}{\rightleftharpoons}}\; CH_3CH_{3(g)} + S$$

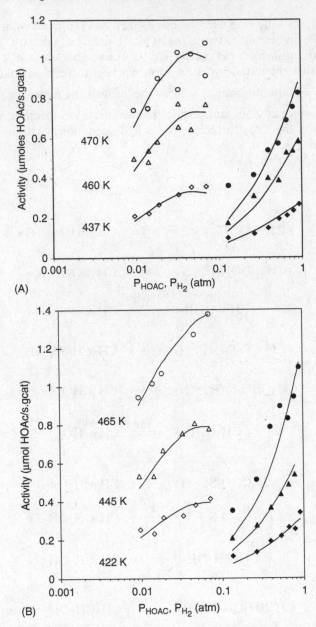

FIGURE 7.15. Activity for acetic acid reduction on Pt/TiO$_2$ catalysts versus acetic acid and hydrogen partial pressures. Solid lines indicate the optimum fits obtained for Eq. 9 and the experimental data points: (A) 0.69% Pt/TiO$_2$ (HTR) at 437, 460 and 470 K; (B) 2.01% Pt/TiO$_2$ (LTR) at 422, 445 and 465K. (Reprinted from ref. 34, copyright © 2002, with permission from Elsevier)

Only the first four reactant adsorption steps are assumed to be quasi-equilibrated (QE); they give H atoms and an acetate (Ac) species on the Pt surface based on the study of Vajo et al. [40], plus adsorbed acetic acid (A) and an acyl (Acy) species on the TiO_2 surface. Steps 7, 10 and 13 are reversible and allow for the respective desorption of acetaldehyde, ethanol and ethane as products. Note that it is implicitly assumed that the ensemble of Ti and O atoms constituting the S sites can simultaneously coordinate a hydroxyl group.

By applying the steady-state approximation (SSA) to the carbonaceous surface intermediates, it is easily shown that the specific rate of disappearance of acetic acid (A) to hydrogenated products is

$$-r_A = r_m = \frac{-1}{m}\frac{dN_A}{dt} = L^*L_S k_1 \theta_{Acy}\theta_H = k\theta_{Acy}\theta_H \qquad (1)$$

where the subscript Acy refers to the acyl species on the titania sites and the subscript H refers to H atoms on the Pt sites. Both site densities along with appropriate probability factors have been assimilated into k. Steps 1-4 are quasi-equilibrated, so

$$\theta_H = K_{H_2}^{1/2}P_{H_2}^{1/2}\theta_*, \qquad (2)$$

$$\theta_{Ac} = K_{Ac}P_A\theta_*^2/\theta_H = K_{Ac}P_A\theta_*/K_{H_2}^{1/2}P_{H_2}^{1/2}, \qquad (3)$$

$$\theta_A = K_A P_A \theta_S, \qquad (4)$$

and

$$\theta_{Acy} = K_{Acy}K_A P_A \theta_S \qquad (5)$$

represent the respective fractional coverages of H atoms, acetate species, acetic acid and acyl species, θ_* is the fraction of empty * sites, and θ_S is the fraction of empty S sites. Substitution of equations 2 and 5 into equation 1 gives:

$$r_m = kK_{Acy}K_A K_{H_2}^{1/2}P_A P_{H_2}^{1/2}\theta_S\theta_* \qquad (6)$$

If the Pt surface is assumed to be nearly saturated with H atoms and acetate species, then θ_H, θ_{Ac}, $\gg \theta_*$ and a balance on * sites yields

$$\theta_* = 1/\left(K_{H_2}^{1/2}P_{H_2}^{1/2} + K_{Ac}P_A/K_{H_2}^{1/2}P_{H_2}^{1/2}\right) \qquad (7)$$

If the sites on titania are assumed to be covered principally by acetic acid molecules and acyl species, consistent with the IR results, then the balance on S sites gives

$$\theta_S = 1/(1 + K_A P_A + K_{Acy}K_A P_A) \qquad (8)$$

Substituting equations 7 and 8 into equation 6 gives the final rate expression:

$$r_m = \frac{kK_{Acy}K_A K_{H_2}^{1/2}P_A P_{H_2}^{1/2}}{\left(K_{H_2}^{1/2}P_{H_2}^{1/2} + K_{Ac}P_A/K_{H_2}^{1/2}P_{H_2}^{1/2}\right)\left(1 + K_A\left(1 + K_{Acy}\right)P_A\right)}$$

$$= \frac{k'P_A P_{H_2}^{1/2}}{\left(K_2 P_{H_2}^{1/2} + K_3 P_A/P_{H_2}^{1/2}\right)\left(1 + K_4 P_A\right)} \tag{9}$$

where $k' = kK_{Acy}K_A K_{H_2}^{1/2}$, $K_2 = K_{H_2}^{1/2}$, $K_3 = K_{Ac}/K_{H_2}^{1/2}$ and $K_4 = K_A\left(1 + K_{Acy}\right)$.

Optimization of the data from the Arrhenius runs and three partial pressure runs for two Pt/TiO_2 catalysts gave the results in Table 7.11, which give the predicted activities in Figure 7.15 and also allow K_{H_2}, K_{Ac} and $K_A\left(1 + K_{Acy}\right)$ to be determined. Plotting these values vs. reciprocal temperature produced the linear relationships shown in Figure 7.16, which gave the enthalpies and entropies of adsorption listed in Table 7.12. The α and β values are derived from the lumped parameters:

$$K_A\left(1 + K_{Acy}\right) = e^{\alpha/R}e^{-\beta/RT} \tag{10}$$

For the limiting case where $\theta_A \gg \theta_{Acy}$, α and β become the entropy and enthalpy for molecular acetic acid adsorption on TiO_2, whereas if $\theta_{Acy} \gg \theta_A$, then these two values represent the dissociative adsorption of acetic acid to form an acyl species on the TiO_2 surface. In either case, α and β along with the enthalpies and entropies of adsorption for H_2 and acetic acid on Pt must satisfy the criteria in Table 6.9, and all of them do. From the results obtained here, the α and β values seem to best represent the dissociative adsorption of acetic acid on TiO_2 to form an acyl group.

The illustration just shown is another example of how a complex sequence of elementary steps reduces to a relatively simple kinetic rate expression after certain assumptions are made; in this case they were that H atoms were the reactive species on Pt while both acetic acid and acyl species existed on titania although the latter appeared to be the MARI on these sites. As

TABLE 7.11. Optimized rate parameters for Pt/TiO_2 catalysts[a] (Reprinted from ref. 34, copyright © 2002, with permission from Elsevier)

Catalyst	$k' \times 10^{-4}$ (μmole/s·g cat·$atm^{3/2}$)	$K_2 \times 10^{-2}$ ($atm^{-1/2}$)	$K_3 \times 10^{-4}$ ($atm^{-1/2}$)	K_4 (atm^{-1})	$K_{H_2} \times 10^{-5}$ (atm^{-1})	$K_{Ac} \times 10^{-7}$ (atm^{-1})
0.69% Pt/TiO_2 (HTR)						
Temp (K) = 437	6.5	22.5	7.5	10.9	50.8	17.0
460	6.8	8.7	5.4	4.4	7.6	4.7
470	8.5	7.0	4.7	5.5	4.8	3.4
2.01% Pt/TiO_2 (LTR)						
Temp (K) = 422	4.7	13.1	6.7	4.3	17.4	8.9
445	6.8	7.8	5.4	3.3	6.1	4.2
465	6.0	2.6	3.6	0.4	0.7	0.9

[a] For equation 9.

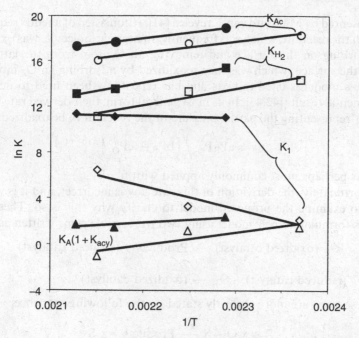

FIGURE 7.16. Arrhenius plots of optimized rate parameters from Table 7.9: 0.69% Pt/TiO$_2$ (HTR) – filled symbols, 2.01% Pt/TiO$_2$ (LTR) – open symbols. (Reprinted from ref. 34, copyright © 2002, with permission from Elsevier)

TABLE 7.12. Enthalpies and entropies of adsorption from rate parameters with 90% confidence limits[a] (Reprinted from ref. 34, copyright © 2002, with permission from Elsevier)

Catalyst	K$_{H_2}$		K$_{Ac}$		K$_A$(1 + K$_{Acy}$)[b]	
	ΔH°_{ad} (kcal/mole)	ΔS°_{ad} (cal/mole ·K)	ΔH°_{ad} (kcal/mole)	ΔS°_{ad} (cal/mole · K)	ΔH° (kcal/mole)	ΔS° (cal/mole ·K)
0.69% Pt/TiO$_2$ (HTR)	-30 ± 14	-37 ± 30	-21 ± 5	-10 ± 12	-10 ± 34	-18 ± 75
2.01% Pt/TiO$_2$ (LTR)	-29 ± 50	-39 ± 114	-20 ± 34	-11 ± 77	-22 ± 72	-47 ± 163

[a] Standard State: 1 atm

Boudart has pointed out in an earlier paper [31]: "Theorem II – In a catalytic sequence of steps, all steps that follow an irreversible step involving the MASI (MARI) as reactant are kinetically not significant."

7.4.2   Redox Reactions: The Mars-van Krevelen Rate Law

Oxidation reactions represent one family of reactions which may very well involve only irreversible steps, especially those reactions conducted at low temperatures. An early model to describe such reactions on an oxide catalyst was presented by Mars and van Krevelen [41]. It consisted of a redox sequence in which the reactant, which was frequently an organic molecule, was oxidized by adsorbing on the surface and removing an O atom from the lattice to reduce the surface, which was then reoxidized by adsorbing an O_2 molecule [41]. This model is cited in texts and has repeatedly been used to describe experimental results [42,43]. In its most general form, their derived rate law is, with P_R representing the partial pressure of the reactant to be oxidized:

$$r_m = aP_R P_{O_2}^n \Big/ \left(bP_R + cP_{O_2}^n\right) \tag{7.30}$$

and it is perhaps most commonly applied with $n = 1$.

Unfortunately, the derivation of this rate law is incorrect, and it is worth-while to examine the proposed model to clarify why this is so. These two authors formulated their model using two irreversible steps written as:

R + (oxidized catalyst) $\longrightarrow$ Products + (reduced catalyst)

(reduced catalyst) + O_2 $\longrightarrow$ (oxidized catalyst)

but these steps are more correctly stated by the following sequence:

(7.26) $R + x\, O - S \xrightarrow{k_1} \text{Products} + x\, S$

(7.27) $x\, S + x/2\, O_2 \xrightarrow{k_2} x\, O - S$

where R represents the reactant and S is an active site (a lattice vacancy).

A first-order dependence on R was assumed, and the rate, step 7.26, was stated to be:

$$r_1 = \beta k_1 P_R \theta \tag{7.31}$$

where β is the number of O_2 molecules $(x/2)$ required and θ is the coverage of the active sites by oxygen. Thus the rate of this first step is clearly inappropriate because it is almost certainly not an elementary step and, in addition, it represents a Rideal-Eley reaction involving multiple bond breaking and multi-body interactions for anything other than the addition of a single O atom. This is typically not the case, although the authors themselves applied it to the oxidation of benzene, toluene, naphthalene and anthracene, which required up to 9 O atoms or more $(x = 9$ in equation 7.26) when products are considered. Thus rate equation 7.31 is incorrect for any value of x other than unity (unless a special site pair is proposed to accommodate an O_2 molecule and allow $x = 2$; however, multiple bond breaking would still be required).

For the second step, the rate of reoxidation (O_2 adsorption) was assumed to be:

$$r_2 = k_2 P_{O_2}^n (1 - \theta) \tag{7.32}$$

This rate is clearly incorrect for any value of n other than unity (see Chapter 5.1), and it is also inappropriate for any type of oxygen adsorption other than molecular adsorption on a single site (or perhaps a vacancy pair defined as an active site). Utilization of the SSA requires that $r_1 = r_2$, and the use of these two incorrect equations (7.31 and 7.32) gives

$$\theta = k_2 P_{O_2}^n \Big/ \left(\beta k_1 P_R + k_2 P_{O_2}^n \right) \qquad (7.33)$$

which yields equation 7.30 after substitution into equation 7.31. It is surprising that this widely-used rate equation has never been thoroughly examined because, as shown here, it can be correctly applied *only* when n = 1, i.e., a very simple two-step sequence without the formation of any by-product. Even then, step 7.26 is inconsistent with step 7.27 if x = 1 because the former does not imply dissociative O_2 adsorption.

It is worth noting at this point that a H-W model invoking product desorption as the RDS could also give equation 7.30 for values of n = 1 or 2, provided the fraction of vacant sites is very low and the fraction of sites covered by the two reactants is very high. It may also be possible to propose a more realistic sequence of steps comprised of reversible and irreversible elementary steps (perhaps even including some quasi-equilibrated steps) that could result in a rate expression of the Mars–van Krevelen form. Consequently, the Mars–van Krevelen rate expression should be considered to be only a mathematical fitting equation with no theoretical basis.

7.5 Data Analysis with an Integral Reactor

The use of differential reactors to simplify the acquisition of accurate rate data has been mentioned previously. Under integral reactor conditions, analysis of the rate data can be much more complicated, even if a RDS is incorporated into the model, as demonstrated in Illustration 7.7.

Illustration 7.7 – The Reduction of NO by CH₄ on La₂O₃ – Integral Reactor Operation and Incorporation of Competitive Product Adsorption into a L-H-type Bimolecular Reaction

The kinetics of this reaction over La_2O_3 and Sr-promoted La_2O_3 have been studied in detail in recent years by Vannice and coworkers [4,44–47]. Most studies were conducted at low conversions under differential reactor conditions; however, with O_2 in the feed, a simultaneous CH_4 combustion reaction can also occur to cause non-differential conversions of CH_4 and O_2 in the catalyst bed, and this must be properly accounted for in analyzing the kinetic results. Furthermore, a previous reaction model had included only reactants in the site balance, but utilization under commercial-type conditions required the consideration of both CO_2 and H_2O in the rate expression. Modification of the previous L-H-type reaction mechanism to include the influence of CO_2 and H_2O demonstrates one of its advantages, i.e., the ease of incorporation of the adsorbates into the model and the rate expression.

Using unsupported La_2O_3, which had been pretreated in O_2 at 1023 K, the kinetics of NO reduction by CH_4 in the presence of O_2 were examined between 773 and 923 K with varying concentrations of O_2, CO_2, and H_2O in the feed stream [45]. BET measurements gave the total surface area of the fresh and used samples, while irreversible NO adsorption at 300 K was employed to estimate the active site density. For example, the fresh La_2O_3 had a surface area of $3.8\,m^2\,g^{-1}$ and the site density was 3.6×10^{18} site m^{-2}, assuming 1 NO molecule per site. This kinetic analysis was complicated not only because integral conversions were attained, but also because two reactions were occurring concurrently. Utilizing the stoichiometries of the two simultaneous reactions, a computer program was developed to determine composition vs. conversion through the catalyst bed and thus allow integration of the reciprocal of the rate (1/r) along with the optimization of the rate parameters to obtain the best fit of the experimental data for NO reduction [48]. Although more complex, such a procedure was necessary for accuracy because CH_4 and O_2 conversions sometimes exceeded 50%. A rate expression determined separately for CH_4 combustion on La_2O_3 was used to describe this reaction [49].

A detailed and rather complex sequence of elementary steps has been proposed to describe NO reduction by CH_4 in the presence of O_2 that is consistent with both NO decomposition and NO reduction by CH_4 in the absence of O_2 as well as with homogeneous free radical chemistry [4]. The details of this catalytic cycle certainly have not been proven; however, as mentioned previously, the assumption of a RDS and the presence of only a limited number of significant surface reaction intermediates simplifies the derived rate expression enormously. With the proposal of a RDS, the catalytic cycle can be reduced to the following steps, with S as an active site, the convention of the arrows denotes whether a step is the RDS or is quasi-equilibrated (See Chapter 2.7), and the stoichiometric number for the elementary step in the cycle is given outside the brackets [47]:

$$(1) \qquad 4[NO + S \overset{K_{NO}}{\rightleftharpoons} NO-S]$$

$$(2) \qquad 2[CH_4 + S \overset{K_{CH_4}}{\rightleftharpoons} CH_4-S]$$

$$(3) \qquad 2[O_2 + 2S \overset{K_{O_2}}{\rightleftharpoons} 2O-S]$$

$$(4) \qquad 3[NO-S + O-S \overset{K_{NO_2}}{\rightleftharpoons} NO_2-S + S]$$

$$(5) \qquad 2[NO_2-S + CH_4-S \overset{k}{\underset{A}{\rightarrow}} HNO_2-S + CH_3-S]$$

$$(6) \qquad \begin{array}{l} 2\,HNO_2-S + 2\,CH_3-S + \\ NO-S + NO_2-S + O-S \end{array} \overset{K}{\rightleftharpoons} \begin{array}{l} 2\,CO_2-S + 4H_2O-S \\ + 2\,N_2 + S \end{array}$$

$$(7) \qquad 2[CO_2-S \overset{1/K_{CO_2}}{\rightleftharpoons} CO_2 + S]$$

$$(8) \qquad 4[H_2O-S \overset{1/K_{H_2O}}{\rightleftharpoons} H_2O + S]$$

$$(9) \qquad 4\,NO + 2\,CH_4 + 2\,O_2 \implies 2\,N_2 + 4\,H_2O + 2\,CO_2$$

The areal rate is:

$$r_a\left(\mu mol\,N_2\,s^{-1}\,m^{-2}\right) = \frac{1}{A}\frac{dN_{N_2}}{dt} = Lk\theta_{NO_2}\theta_{CH_4} \qquad (1)$$

and if the predominant reaction intermediates are assumed to be only adsorbed reactants and products, excluding N_2, i.e., $NO-S$, CH_4-S, $O-S$, CO_2-S and H_2O-S, then the site balance is

$$1 = \theta_v + \theta_{NO} + \theta_{CH_4} + \theta_O + \theta_{CO_2} + \theta_{H_2O} \qquad (2)$$

which, based on steps 1, 2, 3, 7 and 8, gives

$$\theta_v = 1\Big/\left(1 + K_{NO}P_{NO} + K_{CH_4}P_{CH_4} + K_{O_2}^{1/2}P_{O_2}^{1/2} + K_{CO_2}P_{CO_2} + K_{H_2O}P_{H_2O}\right) \qquad (3)$$

With reaction 4,

$$\theta_{NO_2} = K_{NO_2}\theta_{NO}\theta_O/\theta_v \qquad (4)$$

which, after substitution into equation 1 and utilization of quasi-equilibrated steps 1–3 gives

$$r_{N_2} = LkK_{NO_2}K_{NO}K_{CH_4}K_{O_2}^{1/2}P_{NO}P_{CH_4}P_{O_2}^{1/2}\theta_v^2.$$

Combining this with equation 3 produces the final rate law:

$$r_{N_2} = k'P_{NO}P_{CH_4}P_{O_2}^{1/2} \Big/ \Big(1 + K_{NO}P_{NO} + K_{CH_4}P_{CH_4} + K_{O_2}^{1/2}P_{O_2}^{1/2} + K_{CO_2}P_{CO_2}$$

$$+ K_{H_2O}P_{H_2O}\Big)^2 \tag{6}$$

where $k' = LkK_{NO_2}K_{NO}K_{CH_4}K_{O_2}^{1/2}$.

From the stoichiometry of reaction 9, it is clear that the rate of CH_4 consumption for only NO reduction equals the rate of N_2 formation. For the CH_4 combustion reaction

$$(10) \qquad\qquad CH_4 + 2O_2 \Longrightarrow CO_2 + 2H_2O$$

the areal rate equation is [46]:

$$r_{com} = \frac{-1}{A}\frac{dN_{CH_4}}{dt}$$

$$= \frac{k'_{com}P_{CH_4}P_{O_2}^{1/2}}{\Big(1 + K_{NO}P_{NO} + K_{CH_4}P_{CH_4} + K_{O_2}^{1/2}P_{O_2}^{1/2} + K_{CO_2}P_{CO_2} + K_{H_2O}P_{H_2O}\Big)^2} \tag{7}$$

The total rate of CH_4 disappearance (r_{CH_4}) is the sum of equations 6 and 7, i.e.,

$$(-r_{CH_4})_T = (r_{N_2}) + (-r_{com}) \tag{8}$$

Beginning with equation 4.20 and rewriting it in terms of the partial pressure of the limiting reactant, CH_4, one has

$$W/F_{CH_{4_o}} = \int_0^{f_{CH_4}} \frac{df_{CH_4}}{(-r_{CH_4})_T} = \frac{1}{P_{CH_{4_o}}}\int_{P_{CH_{4_o}}}^{P_{CH_4}} \frac{dP_{CH_4}}{\left(-r'_{CH_4}\right)_T} \tag{9}$$

where $P_{CH_{4_o}}$ is the partial pressure of CH_4 at the reactor inlet. In this study the fractional conversion of NO was never greater than 0.1; therefore, it was always differential and an average partial pressure of NO can be used. Next, a selectivity for oxidation to CO_2 vs. CO was defined for reaction 10 to allow for any incomplete combustion to form CO:

$$S_{CO_2} = mole\ CO_2/(mole\ CO_2 + mole\ CO) \tag{10}$$

Consequently, one can write the following relationships:

$$P_{NO} = P_{NO_o}(1 - f_{NO}/2) \tag{11}$$

$$P_{H_2O} = P_{H_2O_0} + 2(P_{CH_{4_0}} - P_{CH_4}) \tag{12}$$

$$P_{CO_2} = P_{CO_{2_0}} + (P_{CH_{4_0}} - P_{CH_4})S_{CO_2} + f_{NO}P_{NO}\left(\frac{1 - S_{CO_2}}{4}\right) \tag{13}$$

$$P_{O_2} = P_{O_{2_0}} - (P_{CH_{4_0}} - P_{CH_4})\left(\frac{3 + S_{CO_2}}{2}\right) + f_{NO}P_{NO}\left(\frac{2 - S_{CO_2}}{4}\right) - \frac{P_{NO_2}}{2} \tag{14}$$

where P_{i_0} represents the initial partial pressure of component i in the feed. Additional experiments showed that no CO formation resulted from reaction 9 and all came directly from CH_4 oxidation via reaction 10. As a result, the consumption of CH_4 in only reaction 10 is

$$r_{com} = (r_{CH_4})_T - r_{N_2} \tag{15}$$

and the rate of CO_2 formation from reaction 10 is

$$r_{CO_2} = r_{com} - r_{CO} \tag{16}$$

where r_{CO} is the rate of CO formation. Based on these relationships, the selectivity to CO_2 can be written as

$$S_{CO_2} = [(r_{CH_4})_T - r_{N_2} - r_{CO}]/(r_{CH_4})_T \tag{17}$$

At the low concentrations used, the volume change can be neglected and, furthermore, pressure drops were minimal. Therefore, the overall rate of CH_4 disappearance can be written in terms of only P_{CH_4} after equations 11–14 are substituted into equations 6 and 7, and equation 9 can be numerically integrated after equation 8 is substituted.

To accomplish this, a data-fitting program (Scientist 2.01, MicroMath Scientific Software) was used to determine the optimal rate parameters. This software uses Powell's algorithm [50], which is a hybrid of the Gauss-Newton and steepest descent methods, to find the best least squares fit. This program also contained a numerical integration function, based on an adaptive quadrature algorithm from another study [51], which was employed to integrate the reciprocal rate law function.

The capability of this rate expression to describe the data is illustrated in Figures 7.17 and 7.18. The optimized rate parameters at various temperatures are listed in Table 7.13, and Arrhenius plots of these values yield the thermodynamic properties in Table 7.14. The enthalpies and entropies of adsorption are consistent with the rules in Table 6.9, considering the uncertainty of the numbers. Again, this does not prove that the proposed model is correct, but only that it is consistent and should not be rejected at this time. This same model was also capable of providing satisfactory fits of the data

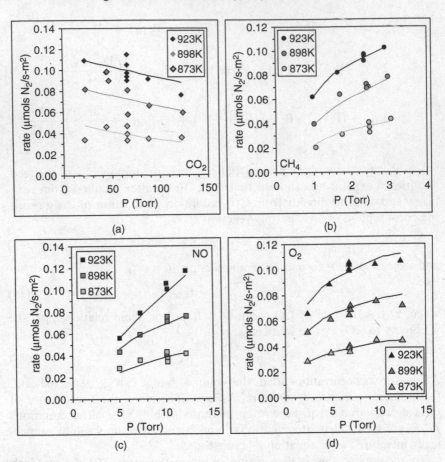

FIGURE 7.17. Partial pressure dependencies for NO reduction by CH_4 in excess CO_2 (and O_2) between 873 and 923K for: (a) CO_2, (b) CH_4, (c) NO, and (d) O_2. Experimental data are represented by symbols and the predicted rates given by Eq. *9*, after substitution of Eqs. *6, 7* and *8*, are represented by lines. Standard reaction conditions: 1.4% NO, 0.35% CH_4, 1% O_2, 9% CO_2, balance He; GHSV $= 1.3 \times 10^5$ h^{-1}. (Reprinted from ref. 45, copyright © 2002, with permission from Elsevier)

and consistent thermodynamic parameters for this reaction on Sr-promotional La_2O_3 [46] and La_2O_3 dispersed on Al_2O_3 [47].

Finally, in all these Examples and Illustrations, it should be emphasized that a complete, balanced catalytic cycle must be used, i.e., when the sequence of steps, elementary or otherwise, is added, all reactive intermediates on both sides must cancel and only the overall reaction of reactants to products remains.

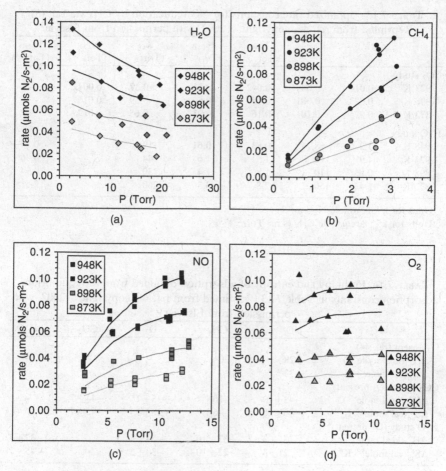

FIGURE 7.18. Partial pressure dependencies for NO reduction by CH_4 in excess H_2O (and O_2) between 873 and 948K for: (a) H_2O, (b) CH_4, (c) NO, and (d) O_2. Experimental data are represented by symbols and the predicted rates given by Eq. 9, after substitution of Eqs. 6, 7 and 8, are represented by lines. Standard reaction conditions: 1.4% NO, 0.35% CH_4, 1% O_2, 2% H_2O, balance He; GHSV $= 1.5 \times 10^5$ h^{-1}. (Reprinted from ref. 45, copyright © 2002, with permission from Elsevier)

7.6   Occurrence of a Very High Reaction Order

Periodically, a rate dependence on a given reactant can have a very high reaction order, especially in hydrogenation reactions, and partial pressure dependencies on H_2 ranging from 2nd order to 4th order have been reported for benzene hydrogenation [20–24], as shown in Illustration 7.4. The probability for an interaction among three surface intermediates to form a single transition state in a RDS is very low, and it decreases precipitously as the

TABLE 7.13. Optimized rate constants for NO reducation by CH_4 via Eq. (8) (Reprinted from ref. 45, copyright © 2002, with permission from Elsevier)

	$(k')^a$	$k'_{com}{}^b$	K_{NO} (Torr^{-1})	K_{CH_4} (Torr^{-1})	K_{O_2} (Torr^{-1})	K_{CO_2} (Torr^{-1})	K_{H_2O} (Torr^{-1})
CO_2 study							
923 K	0.014	0.049	0.042	0.55	0.37	0.0042	–
898 K	0.023	0.081	0.15	0.60	0.59	0.010	–
873 K	0.025	0.101	0.16	0.91	0.65	0.017	–
H_2O study							
948 K	1.2	68	0.83	0.61	19	–	0.35
923 K	1.9	110	1.3	1.6	44	–	0.46
898 K	2.8	110	1.8	1.4	130	–	0.94
873 K	5.4	150	2.8	3.3	260	–	2.5

[a] Units for k' are μmole N_2(s m^2 Torr$^{2.5}$)$^{-1}$.
[b] Units for k'_{com} are μmole CH_4 (s m^2 Torr$^{2.5}$)$^{-1}$.

TABLE 7.14. Enthalpy and entropy of adsorption obtained from the equilibrium adsorption constants in Table 7.13 (Reprinted from ref. 45, copyright © 2002, with permission from Elsevier)

	NO	CH_4	O_2	CO_2	H_2O
O_2 absent (ref. 44)					
ΔH^o_{ad}(kcal mole^{-1})	−28	−20	–	–	–
ΔS^o_{ad}(cal mole^{-1} K^{-1})	−23	−9	–	–	–
CO_2 study (O_2 present)^a					
ΔH^o_{ad}(kcal mole^{-1})	−43±20	−16±7	−35±15	−44±7	–
ΔS^o_{ad}(cal mole^{-1} K^{-1})	−40±20	−6±7	−28±15	−45±7	
H_2O study (O_2 present)^a					
ΔH^o_{ad}(kcal mole^{-1})	−31±3	−32±10	−54±5	–	−37±7
ΔS^o_{ad}(cal mole^{-1} K^{-1})	−21±3	−22±10	−38±5	–	−45±7

[a] With 90% confidence limits

number of surface species in this interaction increases further. One-body and two-body reactions on a surface (not counting the surface itself) are clearly favored statistically; consequently, reaction orders greater than 2 for a single reactant may be difficult to justify. One obvious possibility for reaction orders between 1 and 2 for a given surface intermediate is an interaction between the same reactant in a RDS. Another explanation in reactions involving the overall addition of more than one mole of a reactant in the catalytic cycle, such as the hydrogenation of an aromatic molecule, is the existence of quasi-equilibrated addition steps prior to the RDS. For example, a 3rd-order dependence on H_2 would be obtained if the addition of the last (6th) H atom to the nearly hydrogenated adsorbed benzene molecule were the RDS. In principle, with this approach any reaction order on H_2 between 0 and 3 could be obtained for the hydrogenation of a single aromatic ring (such as benzene, toluene, or xylene), depending on the choice of the RDS.

However, none of these approaches can account for a reaction order greater than 3. What other possibilities might do so? Another situation that could exist which would produce positive orders on a given reaction is that in which one of the reactants forms an inactive species that occupies an active site (an inhibitor) by losing an atom or a molecule of the other reactant. Partially dehydrogenated species are known to exist in hydrogenation reactions under certain conditions, and such species can eventually lead to coke, which acts as a poison. Thus under low H_2 pressures during the hydrogenation of aromatic molecules, for example, hydrogen-deficient species can be created which block active sites, and an additional role of H_2 then becomes that of moderating the surface coverage of these species, i.e., higher H_2 pressures not only enhance the RDS, but they also increase the fraction of available active sites. This has been proposed previously for benzene and toluene hydrogenation on metals [22,27,52–54] and, in principle, such a model can generate very high reaction orders on H_2. For instance, if the surface coverage of hydrogen-deficient species is high, the partial pressure dependency on H_2 can increase up to 1/2 order for each H atom lost from an adsorbed aromatic molecule [27]. Depending on the reaction being studied, if the reaction order of a reactant begins to exceed 1 or 2, one might want to consider whether this, or a similar, model invoking the control of an inhibitive intermediate, might be applicable.

References

1. O. A. Hougen and K. M. Watson, "Chemical Process Principles. Part 3. Kinetics and Catalysis", Wiley, NY, 1947.
2. M. Boudart and G. Djéga-Mariadassou, "Kinetics of Heterogeneous Catalytic Reactions", Princeton Press, Princeton, NJ, 1984.
3. K. J. Laidler, "Catalysis", P. H. Emmett, Ed., Vol. 1, Chap. 4, Reinhold, NY, 1954.
4. M. A. Vannice, A. B. Walters, and X. Zhang, J. Catal. 159 (1996) 119.
5. S. Lacombe, C. Geantet, and C. Mirodatos, J. Catal. 151 (1994) 439.
6. T. Yamashita and A. Vannice, J. Catal. 161 (1996) 254.
7. L. Rheaume and G. Parravano, J. Phys. Chem. 62 (1959) 264.
8. W. E. Garner and T. Ward, J. Chem. Soc. (1939) 837.
9. J. Saito, J. Chem. Soc. Jpn. 72 (1951) 262.
10. K. Tanaka and A. Ozaki, Bull. Chem. Soc. Jpn. 40 (1967) 420.
11. R. M. Rioux and M. A. Vannice, J. Catal. 216 (2003) 362.
12. R. M. Rioux, M.S. Thesis, The Pennsylvania State University, 2001.
13. D. A. Chen and C. M. Friend, Langmuir 14 (1998) 1451.
14. M. K. Weldon and C. M. Friend, Chem. Rev. 96 (1996) 1391.
15. M. A. Vannice, W. Erley and H. Ibach, Surf. Sci. 254 (1991) 1.
16. M. A. Vannice, W. Erley and H. Ibach, Surf. Sci. 254 (1991) 12.
17. E. Shustorovich, Adv. Catal. 37 (1990) 101.

18. J. A. Dumesic, D. F. Rudd, L. M. Aparicio, J. E. Rekoske and A. A. Trevino, "The Microkinetics of Heterogeneous Catalysis", Chap. 5, ACS, Washington, D.C., 1993.
19. T. Yamashita and A. Vannice, *J. Catal.* 163 (1996) 158.
20. H. Kubicka, *J.Catal.* 12 (1968) 223.
21. R. Z. C. Van Meerten and J. W. E. Coenen, *J. Catal.* 46 (1977) 13.
22. P. Chou and M. A. Vannice, *J. Catal.* 107 (1987) 129.
23. R. Badilla-Ohlbaum, H. J. Neuburg, W. F. Graydon and M. J. Phillips, *J. Catal.* 47 (1977) 273.
24. K. J. Yoon and M. A. Vannice, *J. Catal.* 82 (1983) 457.
25. K. J. Yoon, P. L. Walker, L. N. Mulay and M. A. Vannice, *I&EC Prod. Res. Dev.* 22 (1983) 519.
26. K. J. Yoon, Ph.D. Thesis, The Pennsylvania State University, 1982.
27. S. D. Lin and M. A. Vannice, *J. Catal.* 143 (1993) 563.
28. E. K. Rideal, *Proc. Cambr. Phil. Soc.* 35 (1939) 130; *Chem. Ind.* 62 (1943) 335.
29. K. I. Laidler, "Chemical Kinetics", 3rd Ed., Harper & Row, NY, 1987.
30. D. D. Eley and E. K. Rideal, *Proc. Royal Soc. London A* 178 (1941) 429.
31. M. Boudart, *AIChE J.* 18 (1972) 465.
32. J. H. Sinfelt, H. Hurwitz, and R. A. Shulman, *J. Phys. Chem.* 64 (1960) 1559.
33. W. Rachmady and M. A. Vannice, *J. Catal.* 192 (2000) 322.
34. W. Rachmady and M. A. Vannice, *J. Catal.* 207 (2002) 317.
35. W. Rachmady and M. A. Vannice, *J. Catal.* 208 (2002) 158.
36. W. Rachmady and M. A. Vannice, *J. Catal.* 209 (2002) 87.
37. W. Rachmady, Ph.D. Thesis, The Pennsylvania State University, 2001.
38. M. A. Vannice, *J. Molec. Catal.* 59 (1990) 165.
39. M. A. Vannice, *Topics in Catal.* 4 (1997) 241.
40. J. J. Vajo, Y.-K. Sun and W. H. Weinberg, *Appl. Surf. Sci.* 29 (1987) 165.
41. P. Mars and D. W. van Krevelen, *Chem. Eng. Sci. (Special Supplement)* 3 (1954) 41.
42. J. M. Smith, "Chemical Engineering Kinetics", 3rd Ed., McGraw-Hill, NY, 1987.
43. R. I. Masel, "Principles of Adsorption and Reaction on Solid Surfaces", Wiley, NY, 1996.
44. S.-J. Huang, A. B. Walters and M. A. Vannice, *Appl. Catal. B* 17 (1998) 183.
45. T. J. Toops, A. B. Walters and M. A. Vannice, *Appl. Catal. B* 38 (2002) 183.
46. T. J. Toops, A. B. Walters and M. A. Vannice, *Catal. Lett.* 82 (2002) 45.
47. T. J. Toops, A. B. Walters and M. A. Vannice, *J. Catal.* 214 (2003) 292.
48. T. J. Toops, Ph.D. thesis, The Pennsylvania State University, 2001.
49. T. J. Toops, A. B. Walters and M. A. Vannice, *Appl. Catal. A* 233 (2002) 125.
50. M. J. Powell, in "Numerical Methods for Nonlinear Algebraic Equations", P. Robinwitz, Ed., Gordon & Breach Science, NY, 1970.
51. T. N. L. Patterson, "Mathematics of Computation", 28 (1974) 344.
52. P. Chou and M. A. Vannice, *J. Catal.* 107 (1987) 140.
53. S. D. Lin and M. A. Vannice, *J. Catal.* 143 (1993) 539.
54. S. D. Lin and M. A. Vannice, *J. Catal.* 143 (1993) 554.
55. J. H. Sinfelt, *Catal. Rev.* 3 (1970) 175.
56. A. Dandekar and M. A. Vannice, *Appl. Catal. B* 22 (1999) 179.
57. B. Sen and M. A. Vannice, *J. Catal.* 113 (1988) 52.
58. X. Zhang, A. B. Walters and M. A. Vannice, *Appl. Catal. B* 4 (1994) 237; 7 (1996) 321.
59. S.-J. Huang, A. B. Walters and M. A. Vannice, *J. Catal.* 173 (1998) 229.

60. M. C. J. Bradford, P. E. Fanning and M. A. Vannice, *J. Catal.* 172 (1997) 479.
61. M. C. J. Bradford and M. A. Vannice, *Appl. Catal. A* 142 (1996) 97.
62. C.-F. Mao and M. A. Vannice, *J. Catal.* 154 (1995) 230.

Problem 7.1

Derive a rate expression for each of the following single reactions taking place through a sequence of steps as indicated. Define your rate clearly. S represents an active site.

a.
$$O_3 + M \xrightarrow{k_1} O_2 + O + M$$

$$\frac{O + O_3 \xrightarrow{k_2} 2\,O_2}{2\,O_3 \implies 3\,O_2}$$

b.
$$N_2O + S \xrightarrow{k_1} N_2 + S{-}O$$

$$\frac{CO + S{-}O \xrightarrow{k_2} CO_2 + S}{N_2O + CO \implies N_2 + CO_2}$$

c.
$$C_2H_6\,_{(g)} \underset{}{\overset{K}{\rightleftharpoons}} C_2H_2\,_{(ad)} + 2\,H_2\,_{(g)}$$

$$C_2H_2\,_{(ad)} + H_2\,_{(g)} \xrightarrow{k_1} 2\,CH_2\,_{(ad)}$$

$$\frac{2[CH_2\,_{(ad)} + H_2\,_{(g)} \xrightarrow{k_2} CH_4\,_{(g)}]}{C_2H_6 + H_2 \implies 2\,CH_4}$$

Assume single-site adsorption and that $C_2H_{2(ad)}$ is the MARI. This last sequence is shown as originally written in the literature and can be ambiguous. Rewrite it using defined active sites and derive a rate expression.

Problem 7.2

Derive the form of the rate expression based on the elementary steps below, where S is an active site: a) Assuming [B−S] is the MARI, and b) Assuming [A−S] is the MARI.

Step 1
$$A + S \overset{K_1}{\rightleftharpoons} A{-}S$$

Step 2
$$A{-}S \overset{K_2}{\rightleftharpoons} B{-}S$$

Step 3
$$B{-}S \xrightarrow{k_3} B + S$$

$$\frac{}{A \implies B}$$

Problem 7.3

Consider the following sequence of steps to describe the high-temperature kinetics of NO reduction by CH_4 on La_2O_3 and Sr/La_2O_3 in the absence of O_2: (a) Provide any necessary stoichiometric coefficient for each step to balance the overall reaction, i.e., no active centers are contained in the overall stoichiometry; (b) Derive a kinetic rate expression from this sequence assuming NO* and CH_2* are the only two significant surface species; (c) Does the final rate expression have the capability of fitting the rate data reported in the paper by Vannice et al. [4], which gave reaction orders of 0.19–0.26 for CH_4 and 0.73–0.98 for NO? Why?

$$NO + * \underset{}{\overset{K_{NO}}{\rightleftharpoons}} NO*$$

$$NO* + * \underset{}{\overset{K_1}{\rightleftharpoons}} N* + O*$$

$$CH_4 + O* \underset{}{\overset{K_2}{\rightleftharpoons}} CH_2* + H_2O$$

$$CH_2* + NO* \xrightarrow{k} CHNO* + H*$$

$$H* + CHNO* + 2 NO* \underset{}{\overset{K_3}{\rightleftharpoons}} CO_2 + N_2 + H_2O + N* + 3 *$$

$$\underline{2 N* \underset{}{\overset{1/K_{N_2}}{\rightleftharpoons}} N_2 + 2 *}$$

$$4 NO + CH_4 \implies 2 N_2 + CO_2 + 2 H_2O$$

Problem 7.4

A series of elementary steps to describe the water gas shift reaction over copper has been proposed as shown below. Define your reaction rate and derive a rate expression consistent with this model, where * indicates an active site on Cu, assuming that the surface concentration of O atoms is negligible.

$$CO + * \underset{}{\overset{K_1}{\rightleftharpoons}} CO*$$

$$H_2O + * \xrightarrow{k_2} H_2O*$$

$$H_2O* \xrightarrow{k_3} H_2 + O*$$

$$\underline{O* + CO* \xrightarrow{k_4} CO_2 + 2 *}$$

$$CO + H_2O \implies H_2 + CO_2$$

Problem 7.5

Derive the rate expression in Table 7.10:

a) for a bimolecular reaction with nondissociative adsorption as the RDS;
b) for a bimolecular reaction with nondissociative adsorption, but now product desorption is the RDS;
c) for a bimolecular reaction with dissociative adsorption of one reactant as the RDS;
d) for a bimolecular reaction with dissociative adsorption of one reactant and with desorption of the product as the RDS.

Problem 7.6

Sinfelt has studied ethane hydrogenolysis, $C_2H_6 + H_2 \rightarrow 2CH_4$, over the Group VIII metals, and he has proposed the following sequence of steps on cobalt [55]:

$$C_2H_6\,_{(g)} \underset{k_{-1}}{\overset{k_1}{\rightleftharpoons}} C_2H_5\,_{(ad)} + H_{(ad)}$$

$$C_2H_5\,_{(ad)} + H_{(ad)} \overset{K_2}{\rightleftharpoons} C_2H_4\,_{(ad)} + H_{2(g)}$$

$$C_2H_4\,_{(ad)} + H_2\,_{(g)} \overset{k_3}{\longrightarrow} 2\,CH_3\,_{(ad)}$$

$$2\,CH_3\,_{(ad)} + H_2 \overset{K_4}{\rightleftharpoons} 2\,CH_4\,_{(g)}$$

Defining the rate as:

$$r = -d[C_2H_6]/dt = -d[H_2]/dt = Lk_3\theta_{C_2H_4}P_{H_2},$$

(a) derive the rate expression for this reaction, as written, in terms of C_2H_6 and H_2 pressures. This sequence was taken from the paper as written.
(b) Rewrite the sequence to include active sites – note the subtle differences regarding the inclusion of active sites – and derive the rate equation again. What assumption must be made to bring agreement between the two rate expressions?

Problem 7.7

The decomposition of N_2O over a 4.56% Cu/ZSM-5 catalyst (ZSM-5 is a zeolite discovered by Socony Mobil, hence the letter designation) has been investigated by Dandekar and Vannice [56]. Temperatures were varied from 623 to 673 K and partial pressure dependencies were determined for N_2O, O_2 and N_2 at three different temperatures under approximately differential reaction conditions. The rate data are given in Table 1. The N_2

TABLE 7.1. N_2O decomposition activity versus the partial pressures of N_2O and O_2 (from ref 56)

T(K)	P_{N_2O} (atm)	P_{O_2} (atm)	Activity (μ mole s^{-1}g^{-1})
623	0.066	0	0.39
623	0.133	0	0.42
623	0.261	0	0.45
623	0.533	0	0.47
653	0.066	0	1.6
653	0.133	0	2.1
653	0.261	0	2.4
653	0.533	0	2.6
673	0.066	0	5.5
673	0.133	0	7.5
673	0.261	0	8.6
673	0.533	0	9.9
623	0.10	0.01	0.51
623	0.10	0.06	0.27
623	0.10	0.12	0.22
623	0.10	0.24	0.19
653	0.10	0.01	1.92
653	0.10	0.06	1.49
653	0.10	0.12	1.44
653	0.10	0.24	1.33
673	0.10	0.01	6.6
673	0.10	0.06	6.1
673	0.10	0.12	5.7
673	0.10	0.24	5.5

TABLE 2. Optimized rate parameters for L-H rate expression (from ref. 56)

T (K)	Lk (μmole s^{-1}gcat^{-1})	K_{N_2O} (atm^{-1})	K_{O_2} (atm^{-1})
623	0.40	36	9
653	3.00	16	4
673	12.6	13	2

pressure had no effect on the rate. What are the reaction orders obtained from a power rate law? Additional studies showed O_2 desorption to be quite facile at temperatures above 623 K and activity maintenance was very good. Consequently, at these higher temperatures, what is the simplest L-H model describing this reaction? Assume that the only significant reaction intermediates are adsorbed N_2O molecules and O atoms. What is the rate expression? These authors report an apparent activation energy of 36.2 kcal mole^{-1}, and the values for the optimized rate parameters in the rate expression are in Table 2, where k is associated with the RDS and it incorporates L. Are these parameters consistent? Why? What is the activation energy for the RDS? CO adsorption at 300 K was used to count the Cu_s^{+1} cations in the zeolite matrix, as shown in Figure 7.19. What is the irreversible CO uptake? The adsorption stoichiometry for this process can be assumed to be $CO_{ad}/Cu_s^{+1} = 1$ (See

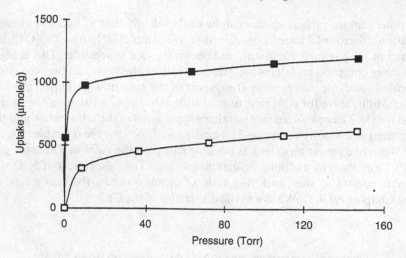

FIGURE 7.19. CO adsorption isotherms for 4.56% Cu/ZSM-5 after pretreatment in He for 1 h at 773K. (Reprinted from ref. 56, copyright © 1999, with permission from Elsevier)

Chapter 3.3.4.2). Calculate the Cu dispersion. What is the N_2O TOF at 823 K and an N_2O partial pressure of 0.0666 atm in the feed under differential reaction conditions?

Problem 7.8

N_2O decomposition has also been investigated on a 4.9% $Cu/\eta - Al_2O_3$ catalyst in which, after a pretreatment in H_2 at 573 K, the Cu was highly dispersed and present almost completely as zero-valent Cu at the surface [56]. These small Cu crystallites exhibited significant deactivation, which was attributed to strong oxygen adsorption at the Cu surface, except at high temperatures above 823 K. Additional studies indeed showed that rapid O_2 desorption required much higher temperatures than with the Cu/ZSM-5 catalysts; therefore, O_2 adsorption/desorption was assumed to be reversible, rather than quasi-equilibrated, and no RDS existed, as shown in the sequence of steps proposed below, where * represents an active site:

(1) $$N_2O + * \underset{}{\overset{K_{N_2O}}{\rightleftharpoons}} N_2O*$$

(2) $$N_2O* \xrightarrow{k} N_2 + O*$$

(3) $$2\,O* \underset{k_{-1}}{\overset{k_1}{\rightleftharpoons}} O_2 + 2\,*$$

Assume that no surface species can be neglected and that k_1 and k_{-1} already contain a factor of 2 (due to $d\xi/dt = dN_i/\nu_i dt$, thus $d[O^*]/dt = 2k_1'[O^*]^2$ for step 3 in the forward direction), and derive the rate expression. This is algebraically complex, and demonstrates how rate expressions can become complicated once one moves away from some of the common simplified models. Suggestion: solve for $[O^*]$ first to substitute into the site balance, then solve that for $[^*]$. The optimized rate parameters are listed in the table below and the apparent activation energy was 34.4 kcal mole^{-1}. Are they reasonable? Evaluate them to as great an extent as possible based on the rules in Table 6.9 and 6.10. (You need to use only the maximum rate). The adsorption of N_2O was used to count Cu_s^0 sites, and 354 μmole 'O' atoms was adsorbed per g catalyst (see Chapter 3.3.4.3). What was the dispersion of the Cu?

Optimized rate parameters for N_2O decomposition (from ref. 56)

T (K)	Lk (μmole s^{-1} g^{-1})	K_{N_2O} (atm^{-1})	Lk$_1$ (μmole s^{-1} g^{-1})	Lk$_{-1}$ (μmole s^{-1} g^{-1})
803	0.18	0.38	0.10	1.1×10^{-4}
823	0.51	0.23	0.14	0.7×10^{-4}
843	1.10	0.18	0.18	0.5×10^{-4}

Problem 7.9

The kinetics of acetone hydrogenation over 5.0% Pt/SiO$_2$ and 1.9% Pt/TiO$_2$ catalysts, previously characterized by H$_2$ chemisorption, were studied by Sen and Vannice [57]. The dispersion of Pt was 0.31 in the former catalyst and in the latter catalyst, which exhibited SMI behavior, $D_{Pt} = 0.75$. The kinetic parameters at 303 K and 1 atm from a power rate law, $TOF_{IPA}(s^{-1}) = Ae^{-E/RT}P_{Ace}^X P_{H_2}^Y$, are given in the table below. Because of the uncertainty in Y, consider it to be unity. Propose a model that yields a derived rate expression consistent with these results knowing that dissociative H$_2$ adsorption occurs on Pt. If more than one reaction model can be proposed which gives acceptable rate expressions, can you evaluate which might be rejected? How? (See Problem 6.7 regarding the heat of adsorption of an on-top ($\eta^1\mu_1$) acetone species vs. that for a di-σ-bonded ($\eta^2\mu_2$) species which has both a C atom and the O atom interacting with a surface metal atom.)

Power law rate parameters for acetone hydrogenation (Reprinted from ref. 57, copyright © 1988, with permission from Elsevier)

Catalyst	TOF$_{IPA}^a$ (s^{-1})	A (s^{-1} atm$^{-(X+Y)}$)	E (kcal mole^{-1})	X	Y
5.0% Pt/SiO$_2$	1.2	$6.0 \times 10^{9*}$	16.0	-0.2 ± 0.1	1.2 ± 0.4
1.9% Pt/TiO$_2$	560	1.9×10^{12}	16.3	-0.6 ± 0.3	1.3 ± 0.5

* Note typographical error in reference 57.

^{a} Average TOF$_{IPA}$ @ 303 K and 1 atm

Problem 7.10

As an alternative to the model proposed in Illustration 7.3, consider instead the following H-W-type reaction mechanism for NO decomposition, in which unimolecular decomposition occurs with an additional active site, *, and N_2 desorption is the rate determining step:

(1) $$2[NO + * \underset{}{\overset{K_{NO}}{\rightleftharpoons}} NO*]$$

(2) $$2[NO* + * \underset{}{\overset{K_1}{\rightleftharpoons}} N* + O*]$$

(3) $$2\,O* \underset{}{\overset{1/K_{O_2}}{\rightleftharpoons}} O_2 + 2\,*$$

(4) $$\underline{\quad 2\,N* \overset{k_2}{\overset{}{\longrightarrow}} N_2 + 2\,* \quad} \qquad \text{(RDS)}$$

$$2\,NO \implies N_2 + O_2$$

The stoichiometric number for steps 1 and 2 in the catalytic cycle is 2, as indicated.

(a) From this sequence of elementary steps, derive the most general form of the rate expression.

What rate law is obtained when each of the following assumptions is made:

(b) If [NO *] is assumed to be very low compared to the other surface species,

(c) If [NO *] is assumed to be the MARI,

(d) If [O *] is assumed to be very small compared to the other surface species,

(e) If [O *] is assumed to be the MARI,

(f) If [N *] is assumed to be the MARI,

(g) If [N *] is not only the MARI, but the surface is also nearly saturated with N atoms.

Over a La_2O_3 catalyst at 923 K, this reaction exhibited a reaction order on NO of about 1.2 with no O_2 in the feed and near 1.5 with O_2 in the feed stream, and the rate decreased significantly with O_2 present [58]. Based on a cursory examination of the rate expressions derived in (b) – (g), can any of them be rejected? Why?

Problem 7.11

Methane combustion on La_2O_3-based catalysts has been studied by Toops et al. [46]. With a 4% Sr-promoted La_2O_3 catalyst ($2.5\,m^2g^{-1}$) operating between 773 and 973 K, $0.5-5$ Torr CH_4 and $3-23$ Torr

O_2 (760 Torr = 1 atm) under differential reactor conditions, an apparent activation energy of 29 kcal mole^{-1} for CO_2 formation was observed. Near 900 K, selectivity to CO_2, rather than CO, was about 75% or higher, and the reaction orders from a power rate law are given in Table 1. Propose a L-H-type model for CO_2 formation with a sequence of elementary steps that results in a derived rate expression consistent with these results. It can be assumed that only the adsorbed reactants and products need be included in the site balance, and dissociative O_2 adsorption occurs. Under low-conversion conditions, the surface concentrations of the products can be ignored, so what is the form of the rate equation? Fitting this latter equation to the data produced the optimized rate parameters listed in Table 2, where k$'$ is the lumped apparent rate constant. Evaluate them to determine if they are consistent and state why.

TABLE 1. Reaction orders for CH_4 combustion on Sr-La$_2$O$_3$ (from ref. 46)

Temperature (K)	CH_4	O_2
873	0.72	0.14
898	0.73	0.15
923	0.72	0.12

TABLE 2. Optimized parameters in L-H-type rate law for CH_4 combustion (from ref. 46)

Temperature (K)	k$'$ (μmole s^{-1} m^{-2} atm$^{-1.5}$)	K_{CH_4} (atm^{-1})	K_{O_2} (atm^{-1})
873	0.34	110	110
898	0.29	68	58
923	0.38	59	44

Problem 7.12

The reduction of NO by H_2 on La$_2$O$_3$ and Sr-promoted La$_2$O$_3$ was examined by Huang et al. [59]. Both N_2 and N_2O were observed as products. The sequence of elementary steps proposed for the catalytic cycle was the following, which invoked one type of site (S) to adsorb and activate H_2 while the second type ($*$) interacted with the oxygen-containing reaction intermediates. Stoichiometric numbers for an elementary step are included when needed:

$$(1) \qquad 2[NO_{(g)} + * \underset{}{\overset{K_{NO}}{\rightleftharpoons}} NO*]$$

$$(2) \qquad 2[H_{2(g)} + 2S \underset{}{\overset{K_{H_2}}{\rightleftharpoons}} 2H-S]$$

$$(3) \qquad NO* + H-S \xrightarrow{k_o} HNO* + S$$

$$(4) \qquad HNO* + NO* \xrightarrow{k_1} N_2O* + OH*$$

$$(5) \qquad N_2O* \xrightarrow{k_2} N_{2(g)} + O*$$

$$(6) \qquad N_2O* \underset{k_{-3}}{\overset{k_3}{\rightleftharpoons}} N_2O_{(g)} + *$$

$$(7) \qquad OH* + H-S \underset{}{\overset{K_4}{\rightleftharpoons}} H_2O_{(g)} + * + S$$

$$(8) \qquad \underline{2 H-S + O* \underset{}{\overset{K_5}{\rightleftharpoons}} H_2O_{(g)} + * + 2 S}$$

$$(9) \qquad 2 NO + 2 H_2 \implies N_2 + 2 H_2O$$

Note that step 6 is just a reversible adsorption/desorption step that allows N_2O to leave the surface as a product as well as to react further, and this step is not part of the catalytic cycle to form N_2. Derive rate laws for the areal rates of a) NO disappearance, b) N_2 formation, and 3) N_2O formation. You may assume that $NO*$ is the MARI on the $*$ sites and that H coverage of the S sites is very low.

Problem 7.13

It was mentioned in Chapter 7.1 that if more than one active site is required in a unimolecular decomposition reaction, the denominator must be raised to a power greater than unity. Now, if a RDS does not exist, the derivation can become more complicated. An example of this is provided by a study of ammonia decomposition on a 4.8% Ru/carbon catalyst with a dispersion near unity [60]. Using a power rate law of the form $r_m = kP_{NH_3}^{\alpha} P_{H_2}^{\beta}$, values of $\alpha = 0.75$ and 0.69 were obtained at 643 and 663 K, respectively, while values of β were -2.0 and -1.6 at 643 and 663 K, respectively. The elementary steps proposed for the catalytic cycle are given below with their stoichiometric numbers, where * represents an active site:

(1)
$$2[NH_3 + * \xrightleftharpoons{K_1} NH_3*]$$

(2)
$$2[NH_3* + * \underset{k_{-2}}{\overset{k_2}{\rightleftharpoons}} NH_2* + H*]$$

(3)
$$2[NH_2* + * \xrightleftharpoons{K_3} NH* + H*]$$

(4)
$$2[NH* + * \xrightleftharpoons{K_4} N* + H*]$$

(5)
$$3[2\,H* \xrightleftharpoons{K_5} H_2 + 2\,*]$$

(6)
$$2\,N* \xrightarrow{k_6} N_2 + 2\,*$$

$$\overline{\qquad\qquad\qquad\qquad\qquad\qquad\qquad}$$

$$2\,NH_3 \Longrightarrow N_2 + 3\,H_2$$

Note that four steps are quasi-equilibrated, step 2 is reversible and step 6 is irreversible. Derive the rate expression for NH_3 disappearance assuming that adsorbed nitrogen (N*) is the MARI. Note that $K_5 = 1/K_{H_2}$.

Problem 7.14

As stated in this chapter, a reaction sequence need not contain a RDS. A study of N_2O reduction with CO over a 4.9% Cu/Al_2O_3 catalyst provides an example of this [56]. This reaction around 523 K was much more rapid than N_2O decomposition, and the following sequence of elementary steps was proposed to describe it, where * is an active site:

$$K_{N_2O}$$

(1) $$N_2O + * \;\rightleftharpoons\; N_2O*$$

$$K_{CO}$$

(2) $$CO + * \;\rightleftharpoons\; CO*$$

$$k_1$$

(3) $$N_2O* \;\longrightarrow\; N_2 + O*$$

$$k_2$$

(4) $$\underline{CO* + O* \;\longrightarrow\; CO_2 + 2 *}$$

$$N_2O + CO \;\Longrightarrow\; N_2 + CO_2$$

(Note: there are some typographical errors in this sequence in reference [56].)

a) Include all reaction intermediates in the site balance and derive the rate expression.

b) Using this rate expression, the optimized rate parameters given in the following table were obtained. What is the activation energy for step 3? What are the enthalpies and entropies of adsorption for N_2O and CO? Are they physically meaningful according to the rules in Table 6.9? Why?

c) Dissociative N_2O adsorption was used to count surface Cu° sites (See Chapter 3.3.4.3), and the Cu dispersion was 0.91. Use the largest rate constant for step 3, convert it to a turnover frequency, and evaluate it according to the criteria in Table 6.10. Is it consistent? Why?

Optimized parameters in rate expression from part (a) (from ref. 56)

T (K)	Lk_1 (μmole s^{-1} g^{-1})	K_{N_2O} (atm^{-1})	K_{CO} (atm^{-1})	d*	L^2k_2
448	16.3	0.91	1.10	0.33	200
473	48.9	0.26	0.36	0.36	449
523	221	0.20	0.13	0.87	15030
573	2100	0.05	0.09	0.15	123000

*$d = (k_1 K_{N_2O})^2 / k_2 K_{CO}$

Problem 7.15

The catalytic reforming of methane with carbon dioxide, i.e.,

$$CH_4 + CO_2 \rightleftharpoons 2\,H_2 + 2\,CO$$

is a complex high-temperature reaction. It has been studied on nickel catalysts between 673 and 823 K by Bradford and Vannice [61], and the following catalytic cycle was proposed, where * represents an active site:

(1)
$$CH_4 + * \underset{k_{-1}}{\overset{k_1}{\rightleftharpoons}} CH_2* + H_2$$

(2)
$$2[CO_2 + * \overset{K_2}{\rightleftharpoons} CO_2*]$$

(3)
$$H_2 + 2* \overset{K_3}{\rightleftharpoons} 2H*$$

(4)
$$2[CO_2* + H* \overset{K_4}{\rightleftharpoons} CO* + OH*]$$

(5)
$$H* + OH* \overset{K_5}{\rightleftharpoons} H_2O + 2*$$

(6)
$$CH_2* + OH* \overset{K_6}{\rightleftharpoons} CH_2O* + H*$$

(7)
$$CH_2O* \overset{k_7}{\longrightarrow} CO* + H_2$$

(8)
$$3[CO* \overset{1/K_8}{\rightleftharpoons} CO + *]$$

(9)
$$CH_4 + 2CO_2 \Longrightarrow H_2 + 3CO + H_2O$$

Evidence shows that the reverse water gas shift (RWGS) reaction

(10)
$$H_2 + CO_2 \rightleftharpoons CO + H_2O$$

is quasi-equilibrated at these temperatures, so it is included in the catalytic cycle to give the overall stoichiometry shown by reaction 9. Note that the RWGS reaction is obtained by adding reactions 2, 3, 4, 5 and 8. Assume that CH_2O* is the MARI and derive the rate expression in terms of methane disappearance. After an optimized fitting of the rate data in that study, a value of $k_7 = 5.35\ \mu$mole s^{-1} g cat^{-1} at 723 K was reported for a 1.2% Ni/TiO$_2$ catalyst which chemisorbed 2.4 μmole H$_2$ g cat^{-1}. Is this value for k_7 reasonable if the activation energy for this step is 38.0 kcal mole^{-1}? Why?

Problem 7.16

As another alternative to the model proposed in Illustration 7.3, consider a sequence with no RDS on the surface and with the formation of an adsorbed (NO)$_2$ species which then decomposes to give adsorbed N$_2$O and oxygen. For example, with only NO and O$_2$ adsorption quasi-equilibrated:

$$2[NO + * \overset{K_{NO}}{\rightleftharpoons} NO*]$$

$$2 \, NO* \overset{k_1}{\longrightarrow} *(NO)_2 *$$

$$*(NO)_2 * \overset{k_2}{\longrightarrow} N_2O* + O*$$

$$N_2O* \overset{k_3}{\longrightarrow} N_2 + O*$$

$$2 \, O* \overset{1/K_{O_2}}{\rightleftharpoons} O_2 + 2 *$$

$$\overline{\qquad\qquad\qquad\qquad\qquad\qquad\qquad}$$

$$2 \, NO \implies N_2 + O_2$$

(a) Assume that the surface concentrations of adsorbed $(NO)_2$ and N_2O are negligible compared to the other species and derive the rate expression for N_2 formation. (b) What difficulty occurs if these two species are not ignored?

Problem 7.17

Examine the reaction kinetics in Illustration 7.5 again. Focus in particular on steps *10–14*. If a H-W-type model is chosen that assumes irreversible toluene desorption as the RDS, is the derived rate expression consistent with the data? Why?

Problem 7.18

In a study of formaldehyde oxidation over Ag catalysts, the power rate law over Ag powder was $r = kP_{O_2}$ and over supported Ag it was $r = kP_{O_2}^{0.3}P_{H_2CO}^{0.3}$ [62]. The following sequence of steps was proposed:

$$O_{2(g)} + 2 \, S \overset{k_1}{\longrightarrow} 2 \, O-S$$

$$H_2CO_{(g)} + 2 \, O-S \overset{K_2}{\rightleftharpoons} HCO_2-S + OH-S$$

$$HCO_2-S + O-S \overset{k_2}{\longrightarrow} CO_2 + OH-S + S$$

$$2 \, OH-S \overset{1/K_{H_2O}}{\longrightarrow} H_2O_{(g)} + O-S + S$$

$$\overline{\qquad\qquad\qquad\qquad\qquad\qquad\qquad}$$

$$H_2CO + O_2 \implies CO_2 + H_2O$$

Derive a rate expression based on this model. You may assume the OH-S concentration is very small. Can this rate law be further simplified to give either of the two power rate law expressions above? If so, what assumptions are involved in each case?

8
Modeling Reactions on Nonuniform (Nonideal) Surfaces

Real surfaces are known to be nonuniform, so two obvious questions are the following. Can rate expressions for catalytic reactions on nonuniform surfaces be meaningfully and readily derived? If so, do these rate laws more accurately describe the experimental results obtained with real catalysts? The previous chapter was devoted to modeling reactions occurring on a uniform or ideal surface in the Langmurian sense, and the reasons for the validity and success of this approach were cited. The paradox of this successful application of Langmurian types of rate laws to nonideal surfaces was addressed in Chapter 6.5, and the earliest rationale for this observation is still the best explanation, i.e., the reaction proceeding on the most active sites, which may constitute but a small fraction of the total number of sites, dominates the macroscopic kinetic behavior.

8.1   Initial Models of a Nonuniform Surface

The first treatment of the influence of surface nonuniformity on kinetic behavior was that of Constable [1]. He considered the surface to be composed of different sites of the i-th type, with n_i representing the number of sites of type i with an activation energy of E_i. The overall rate constant is then, if the preexponential factor is assumed invariant:

$$k = A \sum_i n_i e^{-E_i/RT} \qquad (8.1)$$

Now, if a distribution function similar to that used in deriving the Freundlich isotherm is proposed (see Chapter 5.4.1), i.e., one invoking an exponential dependence:

$$n_i = a e^{E_i/b}, \qquad (8.2)$$

and Constable gives theoretical reasons to justify this, and if a continuous distribution of sites is then assumed, one has

$$k = Aa \int_{E_1}^{E_2} e^{(1/b - 1/RT)E} dE \qquad (8.3)$$

The lower limit refers to the most active sites and the upper limit refers to the least active sites. Integration of this expression gives:

$$k = \frac{Aa}{(1/b - 1/RT)} \left[e^{(1/b - 1/RT)E_2} - e^{(1/b - 1/RT)E_1} \right] \qquad (8.4)$$

Because $E_2 > E_1$ and the term $(1/b - 1/RT)$ was found experimentally to be negative, the first term in the brackets in equation 8.4, i.e., the upper limit, can be ignored and it simplifies to:

$$k = \frac{-Aa}{(1/b - 1/RT)} e^{(1/b - 1/RT)E_1} = k' e^{E_1/b} e^{-E_1/RT} \qquad (8.5)$$

Consequently, this model of a nonuniform surface not only showed that the most active sites dominate the observed kinetics, but it also provided an explanation for the compensation effect that was periodically observed between the preexponential factor and the activation energy in a given reaction.

8.2 Correlations in Kinetics

There are several relationships in homogeneous kinetics that are very useful, not only in these systems, but also in heterogeneous catalytic reactions [2,3]. These include the Polanyi relation [4,5], which states that for a family of exothermic elementary reactions, the steps are correlated by the expression

$$E_a = E_o + \alpha \Delta H \qquad (8.6)$$

where E_a is the observed activation energy, ΔH is the enthalpy of reaction, E_o is a constant for a given family of reactions, and α is a constant between 0 and 1. This semi-empirical equation is especially useful because it provides a relationship between a known thermodynamic property and a kinetic parameter.

The next is the Brønsted relation, which was originally observed in acid-catalyzed organic reactions in the liquid phase [6], but which also exists for base-catalyzed reactions [7]. It provides a relationship between the rate constant for an acid-catalyzed reaction, k_A, and the ionic strength of the acid as represented by its dissociation constant, K_A:

$$k_A = C K_A^\alpha \qquad (8.7)$$

where C is a constant for a given reaction and α is a transfer constant between 0 and 1. The Brønsted relation can be shown to be a consequence of the Polanyi relation [2,3].

Finally, similar correlations exist in physical organic chemistry that are known as linear free-energy relationships. One of the better known is the Hammett relation [8]:

$$\log(k/k_o) = \rho_k \sigma \tag{8.8}$$

which allows the prediction of a rate constant, k, for a given reaction involving a molecule with a particular substituent group, provided that k_o for a reference reaction and the constant ρ_k for that reaction have been determined and the σ value for that particular substituent is known. A number of values of ρ_k are known, and many values of σ have been tabulated [9]. With these relationships providing tools that link unknown kinetic parameters with measurable thermodynamic properties at our disposal, nonideal or nonuniform surfaces will now be addressed.

8.3   Formalism of a Temkin Surface

The approach taken here is drawn heavily on that of Boudart [2,3,10,11], which is based on the work of Temkin [12–14]. It begins by assuming the reaction can be expressed as a two-step catalytic cycle on a uniform surface of the form

$$(8.1) \qquad A_1 + S_1 \underset{k_{-1}}{\overset{k_1}{\rightleftharpoons}} B_1 + S_2$$

$$(8.2) \qquad S_2 + A_2 \underset{k_{-2}}{\overset{k_2}{\rightleftharpoons}} S_1 + B_2$$

_____

$$(8.3) \qquad A_1 + A_2 \rightleftharpoons B_1 + B_2$$

This is a closed sequence representing a maximum of two reactants and two products, with S_1 representing an empty active site and S_2 representing a filled active site. Application of the SSA (steady-state approximation) to S_1 and S_2 yields for the net rate on these sites:

$$r = r_1 - r_{-1} = L\left[\frac{k_1 k_2 [A_1][A_2] - k_{-1} k_{-2}[B_1][B_2]}{k_1[A_1] + k_{-2}[B_2] + k_2[A_2] + k_{-1}[B_1]}\right] = LN \tag{8.9}$$

where the quantity within the brackets represents a TOF, $N(s^{-1})$.

For this ideal surface the site balance is

$$L = [S_1] + [S_2] \tag{8.10}$$

and at steady state, from the SSA, the ratio of empty to filled sites, u, is

$$u = [S_1]/[S_2] = (k_2[A_2] + k_{-1}[B_1])/(k_1[A_1] + k_{-2}[B_2]) \tag{8.11}$$

The model for a nonuniform surface will be similar to that proposed by Constable, i.e., it is a collection of ensembles of sites, E_i, with identical thermodynamic and kinetic properties within each ensemble. If each property exhibits only very small continuous changes among the ensembles, then, if each ensemble contains dS_j' sites per cm^2, the total site density, L, can be obtained by integration

$$L = \sum_j dS_j' = \int dS_j' \qquad (8.12)$$

and the overall, or total, rate is

$$r_t = \int N_j dS_j' \qquad (8.13)$$

Temkin first hypothesized that a continuous distribution function existed to relate the number of sites to a set of properties, and it had the form

$$dS' = ae^{-\gamma A^\circ/RT} d(A^\circ/RT) \qquad (8.14)$$

where dS' is the site density having a standard affinity for adsorption, A°, between A° and $A^\circ + dA^\circ$ in the ensemble, a is a normalization constant determined by equation 8.12, and γ is a dimensionless parameter characteristic of the surface. It will be shown later that γ corresponds to the γ in the Freundlich isotherm. The affinity is just the negative of the Gibbs free energy ($A^\circ = -\Delta G^\circ$), thus it is positive for a spontaneous, favorable reaction, such as an elementary adsorption step. This will be the parameter used to distinguish different sites.

To relate the rate constants in steps 8.1 and 8.2 to the affinity, which is the property characterizing the sites, Temkin next hypothesized that a Brønsted-type relation existed between the rate constant, k_i, and the equilibrium constant, $K_i = k_i/k_{-i}$, for each elementary step, i.e.,

$$k_i = c_i K_i^{\alpha} \qquad (8.15)$$

where α is a transfer coefficient which lies between 0 and 1, and it is frequently near 1/2 [2,3]. At this point Temkin made the assumption that α was the same for both adsorption and desorption; consequently, using equation 8.15

$$k_1 = c_1 e^{\alpha A_1^\circ/RT} \qquad (8.16)$$

and

$$k_{-1} = k_1/K_1 = c_1 e^{(\alpha-1)A_1^\circ/RT} \qquad (8.17)$$

Similarly for step 8.2, but in the reverse, or adsorption, direction:

$$k_{-2} = c_2 e^{\alpha A_2^\circ/RT} \qquad (8.18)$$

and

$$k_2 = c_2 e^{(\alpha - 1)A_2^o/RT} \qquad (8.19)$$

The affinity of the overall reaction, A_T^o, is a calculable thermodynamic property which is independent of the type of site, and it is related to A_1^o and A_2^o by

$$A_1^o - A_2^o = A_T^o \qquad (8.20)$$

Consequently, there is only one unknown variable involved, which will be chosen to be A_1^o and then redefined, i.e.,

$$A_1^o/RT = A^o/RT = t \qquad (8.21)$$

where t is now a dimensionless affinity characterizing the surface nonuniformity through the distribution function (Eq. 8.14). If t_o and t_1 represent the highest and lowest respective values of t, then the interval

$$f = t_o - t_1 \qquad (8.22)$$

is a measure of the width of nonuniformity. With this definition, the rate constants can be rewritten in terms of t:

$$k_1 = c_1 e^{\alpha t} = c_1 e^{\alpha t_o} e^{\alpha(t - t_o)} = k_1^o e^{\alpha(t - t_o)} \qquad (8.23)$$

and, in similar fashion

$$k_{-1} = k_{-1}^o e^{(\alpha - 1)(t - t_o)} \qquad (8.24)$$

$$k_2 = k_2^o e^{(\alpha - 1)(t - t_o)} \qquad (8.25)$$

$$k_{-2} = k_{-2}^o e^{\alpha(t - t_o)} \qquad (8.26)$$

Note that the superscript ^{o} denotes a value of a preexponential factor corresponding to a rate constant associated with the maximum value of the affinity (t_o).

The normalization constant a can now be evaluated. Equation 8.21 is substituted into equation 8.14, which is then integrated with respect to t as indicated by equation 8.12, i.e.,

$$L = \int_{t_1}^{t_o} a e^{-\gamma t} dt = a\left[-\frac{1}{\gamma}e^{-\gamma t}\right]_{t_1}^{t_o} \qquad (8.27)$$

and

$$a = L\gamma e^{\gamma t_o}/(e^{\gamma f} - 1) \qquad (8.28)$$

This value of a can be placed in equation 8.14, and the fraction of sites with an affinity between t and $t + dt$, dS, can now be expressed:

$$dS = dS'/L = \frac{\gamma e^{\gamma t_o} e^{-\gamma t} dt}{(e^{\gamma f} - 1)} \qquad (8.29)$$

To get the total, or overall, rate on this surface, equation 8.13 must be integrated and, to facilitate this an auxiliary variable, u, as defined by equation 8.11, will be utilized. Thus

$$u = [S_1]/[S_2] = u_o e^{(t_o-t)} \tag{8.30}$$

with

$$u_o = (k_2^o[A_2] + k_{-1}^o[B_1])/(k_1^o[A_1] + k_{-2}^o[B_2]); \tag{8.31}$$

in other words, u_o represents the empty to filled ratio for the sites with the highest affinity. Combining equations 8.9, 8.14, 8.28 and 8.30 into equation 8.13, recognizing that $du = -udt$, and defining $m = \alpha - \gamma$, one obtains [3]:

$$r_t = - \left(\frac{L\gamma}{e^{\gamma f} - 1}\right) \left(\frac{k_1^o k_2^o[A_1][A_2] - k_{-1}^o k_{-2}^o[B_1][B_2]}{(k_1^o[A_1] + k_{-2}^o[B_2])^m (k_2^o[A_2] + k_{-1}^o[B_1])^{1-m}}\right)$$

$$\int_{u_1}^{u_o} \frac{u^{-m}du}{(1+u)} \tag{8.32}$$

This expression can be integrated analytically only with lower and upper limits of 0 and infinity, respectively. However, Temkin assumed that on sites with the lowest affinity (t_1), the coverage would be extremely low and $u \to \infty$, while on sites with the highest affinity (t_o), the coverage would approach saturation and $u \to 0$. Changing the negative sign and the limits on the integral gives

$$\int_o^\infty \frac{u^{-m}du}{1+u} \cong \pi / \sin(\pi m) \tag{8.33}$$

Thus the rate on a nonuniform surface described in this manner is:

$$r_t = L\tau \left(\frac{k_1^o k_2^o[A_1][A_2] - k_{-1}^o k_{-2}^o[B_1][B_2]}{(k_1^o[A_1] + k_{-2}^o[B_2])^m (k_{-1}^o[B_1] + k_2^o[A_2])^{1-m}}\right) \tag{8.34}$$

where $m = \alpha - \gamma$ and

$$\tau = \pi\gamma/(e^{\gamma f} - 1)\sin(\pi m) \tag{8.35}$$

Note the similarities between the rate expression defined by equation 8.34 and that given by equation 8.9 for a uniform surface: the numerators are identical and the denominators contain the same terms, although represented in different ways.

.

8.4 Consequences of Temkin's Model

This derivation was based on a two-step reaction sequence or any reaction that can be simplified to such a sequence. Let us examine how the assumptions involved impact upon the behavior of this surface related to adsorption, kinetic, and catalytic processes.

8.4.1   Adsorption Isotherms

Again consider this nonuniform surface to be comprised of ensembles of sites with each ensemble possessing identical properties. This process for single-site adsorption is described by:

$$(8.4) \qquad A + S_1 \underset{k_{-1}}{\overset{k_1}{\rightleftharpoons}} S_2$$

Within each ensemble, the fraction of occupied sites is

$$[S_2]/([S_1] + [S_2]) \qquad (8.36)$$

thus, with dS as described by equation 8.29 representing the fraction of sites in a given ensemble, the fraction of the total surface that is covered and would be measured macroscopically is

$$\theta = \int_{\text{state } 1}^{\text{state o}} \frac{[S_2]dS}{[S_1] + [S_2]} \qquad (8.37)$$

where $dS = dS'/L$ and dS' is defined by the exponential relationship given in equation 8.14.

Using the adsorption equilibrium defined by step 8.4, $K_1 = k_1/k_{-1} = [S_2]/[A][S_1]$, along with the definition of u in equation 8.30, one gets

$$u = [S_1]/[S_2] = k_{-1}/k_1[A] = u_o/e^{(t-t_o)} \qquad (8.38)$$

and for the sites with the highest affinity, $t = t_o$,

$$u = u_o = k_{-1}^o/k_1^o[A] = 1/K_1^o[A] \qquad (8.39)$$

Substituting equation 8.29 and

$$[S_2]/([S_1] + [S_2]) = \frac{1}{1+u} = \frac{K_1^o[A]e^{(t-t_o)}}{1 + K_1^o[A]e^{(t-t_o)}} \qquad (8.40)$$

into equation 8.37 and noting that $u_1 = u_o e^{(t_o - t_1)} = u_o e^f$, one gets

$$\theta = \frac{\gamma(K_1^o[A])^\gamma}{e^{\gamma f} - 1} \int_{u_o}^{u_1} \frac{u^{\gamma-1}du}{1+u} \qquad (8.41)$$

As discussed in the previous section, $u_o \to 0$ and $u_1 \to \infty$, thus the integral is equal to $\pi/\sin(\gamma\pi)$ and

$$\theta = \left[\frac{\pi\gamma(K_1^o)^\gamma}{(e^{\gamma f} - 1)\sin(\gamma\pi)} \right] [A]^\gamma = \text{constant} \cdot [A]^\gamma \qquad (8.42)$$

This is the empirical form of the Freundlich isotherm, as discussed in Chapter 5.4.1.

Alternatively, if $\gamma = 0$ in equation 8.14, then

$$dS' = a d(A^\circ/RT) = a' dA^\circ \tag{8.43}$$

$$\text{and}\quad \Delta S' = S_1' - S_0' = a'\left(A_1^\circ - A_0^\circ\right) \tag{8.44}$$

$$\text{thus}\quad A_0^\circ = A_1'^\circ - S'/a' = A_1'^\circ - b\theta \tag{8.45}$$

indicating that the standard affinity of adsorption decreases linearly with S' and, because a' is proportional to L (see equation 8.28), this can also be expressed in terms of $\theta = S'/L$. This can also be shown by beginning with equation 8.41 and recognizing that as γ approaches zero,

$$\lim_{\gamma \to 0} \frac{\gamma}{e^{\gamma f} - 1} = 1/f \tag{8.46}$$

and [3]:

$$\theta = 1/f \int_{u_o}^{u_1} \frac{du}{(1+u)u} = 1 - \frac{1}{f}\ln\left(\frac{1+u_o e^f}{1+u_o}\right) = 1 - \frac{1}{f}\ln\left(\frac{K_1^\circ[A] + e^f}{1 + K_1^\circ[A]}\right) \tag{8.47}$$

If the distribution is wide so that $e^f \gg K_1^\circ[A]$, then

$$\theta = \frac{1}{f}\ln\left(1 + K_1^\circ[A]\right) \tag{8.48}$$

and if this relationship is restricted to a region where A is relatively strongly adsorbed and the pressure is not too low, then $K_1^\circ[A] \gg 1$ and θ is:

$$\theta = 1/f \ln K_1^\circ[A]. \tag{8.49}$$

Thus the Temkin isotherm is derived (See Chapter 5.4.2). If the entropy of adsorption does not vary markedly with coverage, which is a reasonable approximation over a wide range of coverage, then equations 8.44 and 8.45 indicate that a linear relationship would exist between the enthalpy of adsorption and the coverage, which is a correlation that has frequently been reported (See Chapter 5.4.2 and Table 5.2).

8.4.2   Kinetic and Catalytic Behavior

To determine what conclusions might be reached about the behavior of a catalyst, let us return to a uniform surface and again consider the two-step reaction described by steps 8.1 and 8.2 in the previous section. The activity per site, N, i.e., the TOF, is given by equation 8.9. If it is now assumed that a Brønsted relation exists for each step, then

$$k_1 = C_1 K_1^\alpha, \; k_{-1} = C_1 K_1^{\alpha-1}, \; k_{-2} = C_2 K_2^\alpha \text{ and } k_2 = C_2 K_2^{\alpha-1}$$

and, because α is frequently near $1/2$, we will allow $\alpha = 1/2$ for simplicity, then the substitution of these values into the expression for N gives

$$N = \frac{C_1 C_2 \left(K^{1/2}[A_1][A_2] - K^{-1/2}[B_1][B_2]\right)}{C_1 K_1^{1/2}[A_1] + C_2 K_2^{1/2}[B_2] + C_2 K_2^{-1/2}[A_2] + C_1 K_1^{-1/2}[B_1]} \qquad (8.50)$$

where K is the equilibrium constant for the overall reaction, i.e., $K = K_1 K_2^{-1}[2]$. If the entropy terms in C_1 and C_2 are assumed to be invariant, then C_1 and C_2 are constant and the numerator is independent of the nature of the catalyst; consequently, the highest TOF, N_{max}, is obtained when the denominator, D, is minimized. For convenience, let us define $X = K_1^{1/2}$ and then determine $dD/dX = 0$ where D is now

$$D = \left(C_1[A_1] + C_2 K^{-1/2}[B_2]\right)X + \left(C_1[B_1] + C_2 K^{1/2}[A_2]\right)X^{-1} \qquad (8.51)$$

The derivative gives the relationship

$$k_1[A_1] + k_{-2}[B_2] = k_{-1}[B_1] + k_2[A_2] \qquad (8.52)$$

which, when substituted into equation 8.11 shows that

$$u = [S_1]/[S_2] = 1 \qquad (8.53)$$

In other words, the optimum catalyst with N_{max} is one with half the sites filled and half empty. This is a quantitative verification of Sabatier's Principle, which states that the best catalyst is one that forms an "unstable intermediate compound" at the surface which is neither too weakly nor too strongly adsorbed [2,3].

There is another important conclusion provided by these results, i.e., the best catalyst for a reaction in the forward direction should not be expected to be the *best* catalyst in the reverse direction. For the former situation far from equilibrium, $[B_1] = [B_2] \cong 0$, and for N_{max}

$$k_1[A_1] = k_2[A_2] \qquad (8.54)$$

However, for the latter situation, again far from equilibrium where $[A_1] = [A_2] \cong 0$, for N_{max}

$$k_{-1}[B_1] = k_{-2}[B_2] \qquad (8.55)$$

The probability that equations 8.54 and 8.55 will be satisfied simultaneously is clearly extremely low, thus verifying this second conclusion.

Let us compare this insight gained about uniform surfaces with conclusions that can be reached regarding the catalytic behavior of nonuniform surfaces. First of all, it should be stated that the concept of a RDS is still applicable to a nonuniform surface [3]. Next, we return to equation 8.9 for a uniform surface, and nonuniformity is introduced by again using a Brønsted relation, but now the affinity is allowed to vary as described by equations 8.21 and 8.22 to give (with $\alpha = 1/2$ for convenience):

$$k_i = c_i K_i^{\alpha} \qquad (8.56)$$

thus $k_1 = k_1^o e^{1/2(t-t_o)}, k_{-1} = k_{-1}^o e^{-1/2(t-t_o)}, k_2 = k_2^o e^{-1/2(t-t_o)}$ and $k_{-2} = k_{-2}^o e^{1/2(t-t_o)}$.
Substituting these terms into the TOF defined by equation 8.9 results in

$$N = \frac{k_1^o k_2^o [A_1][A_2] - k_{-1}^o k_{-2}^o [B_1][B_2]}{(k_1^o[A_1] + k_{-2}^o[B_2])e^{1/2[t-t_o]} + (k_2^o[A_2] + k_{-1}^o[B_1])e^{-1/2(t-t_o)}} \qquad (8.57)$$

With this expression, the numerator is independent of the dimensionless affinity, t, thus N_{max} is again obtained by taking the derivative of the denominator, D, and setting it equal to zero, i.e., $dD/dt = 0$. This gives [3]:

$$(k_2^o[A_2] + k_{-1}^o[B_1])/(k_1^o[A_1] + k_{-2}^o[B_2]) = e^{(t_{max}-t_o)} \qquad (8.58)$$

Taking the square root of both sides and substituting into equation 8.57 provides the maximum reversible rate

$$N_{max} = (\vec{r}_o - \overleftarrow{r}_o)/2(k_1^o[A_1] + k_{-2}^o[B_2])^{1/2}(k_2^o[A_2] + k_{-1}^o[B_1])^{1/2} \qquad (8.59)$$

Note that equation 8.31 and 8.58 are equal, therefore

$$u_o = e^{(t_{max}-t_o)} \qquad (8.60)$$

and from equation 8.30 for any value of t, which is t_{max} in this case,

$$u_{max} = u_o e^{(t_o - t_{max})} = [S_1]/[S_2] = 1 \qquad (8.61)$$

Thus, we see that on the best sites on a nonuniform surface, that is, those with the highest N, the fractional coverage is 1/2 and the number of filled sites equals the number of empty sites, which again is quantitative support for Sabatier's Principle.

It is instructive to examine how N varies with t as t changes on either side of t_{max} (where $N = N_{max}$), so let $t = t_{max} \pm t_i$. Substituting either $t_{max} + t_i$ or $t_{max} - t_i$ into equation 8.57 and using equation 8.58 results in

$$N_i = (\vec{r}_o - \overleftarrow{r}_o)/[(k_1^o[A_1] + k_{-2}^o[B_2])(k_2^o[A_2] + k_{-1}^o[B_1])]^{1/2}\left(e^{+t_i/2} + e^{-t_i/2}\right)$$
$$(8.62)$$

Consequently, one sees that N_i changes symmetrically with t_i on either side of t_{max}, as shown in Figure 8.1.

Furthermore, with this symmetry N_{max} lies at the center of the distribution of dimensionless affinity, $t_o - t_1 = f$, and the lowest value of N (N_{min}) on either side is the same, i.e., that at

$$t_i = \frac{t_o - t_1}{2} = f/2 \qquad (8.63)$$

Substitution of this value of t_i into equation 8.62 gives:

$$N_{min} = (\vec{r}_o - \overleftarrow{r}_o)/[(k_1^o[A_1] + k_{-2}^o[B_2])(k_2^o[A_2] + k_{-1}^o[B_1])]^{1/2}(e^{f/4}) \qquad (8.64)$$

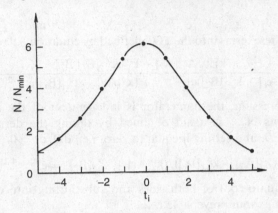

FIGURE 8.1. Variation of rate, indicated by a ratio of turnover numbers, over a range of values of t determined by f = 10. A volcano plot results (Ref. 3).

because the term $e^{-f/4}$ can invariably be neglected relative to $e^{f/4}$. A comparison between the maximum and minimum TOFs on this nonuniform surface is obtained by comparing equations 8.59 and 8.64, which shows that

$$N_{max}/N_{min} = e^{f/4}/2 \qquad (8.65)$$

In summary, the activity of a collection of sites comprising a nonuniform surface has been shown to change because of three factors. The first is thermodynamic and is represented by the variation in the affinity for adsorption ($A^{\circ}/RT = t$) such that the highest and lowest values of the equilibrium adsorption constant are

$$K_{t_o}/K_{t_i} = e^{t_o - t_i} = e^{f} \qquad (8.66)$$

The second factor is kinetic, and the effect on the rate constants is defined by a Brønsted relation which is, using $\alpha = \frac{1}{2}$ for simplicity in equation 8.23:

$$k_{t_o}/k_{t_i} = e^{f/2} \qquad (8.67)$$

Finally, the third factor is the resultant catalytic effect described by equation 8.65, which is a consequence of a compensation effect that alters k_i and K_i in the same direction. Thus as sites become better (more active), there are fewer of them present for reaction. The net result is that a nonuniform surface appears to behave catalytically much more similarly to a uniform surface than expected based on its thermodynamic properties, and it provides justification for the common usage of L-H-type and H-W-type rate expressions.

There have been some very successful applications of rate equation 8.34 derived for a nonuniform surface. One of the best examples is the rate equation of Temkin and Pyzhev which describes the ammonia synthesis reaction [15], and it is discussed in Illustration 8.1.

Illustration 8.1 – Reactions on a Nonuniform Surface – The Ammonia Synthesis Reaction

The well-known experimental rate expression for ammonia synthesis on a doubly promoted iron catalyst was obtained by Temkin and Pyzhev in 1940 [15], i.e.,

$$r = \vec{r} - \overleftarrow{r} = \vec{k}\,[N_2]([H_2]^3/[NH_3]^2)^m - \overleftarrow{k}\,([NH_3]^2/[H_2]^3)^n \qquad (1)$$

where, experimentally, m and n ranged from 1/2 to 2/3 but were frequently near 1/2. The following two-step reaction sequence was proposed:

(1)
$$N_2 + * \underset{k_{-1}}{\overset{k_1}{\rightleftharpoons}} N_2*$$

(2)
$$N_2* + 3\,H_2 \overset{K_2}{\rightleftharpoons} 2\,NH_3 + *$$

(3)
$$N_2 + 3\,H_2 \rightleftharpoons 2\,NH_3$$

where step 1 is a reversible RDS, N_2* is assumed to be the MARI, and step 2 is quasi-equilibrated. With these stipulations, k_2^o, $k_{-2}^o \gg k_1^o$, k_{-1}^o. By analogy with reactions 8.1 and 8.2 (with $[B_1] = 1$), substitution into equation 8.34 for a nonuniform surface gives:

$$r = L\tau\left(\frac{k_1^o k_2^o [N_2][H_2]^3 - k_{-1}^o k_{-2}^o [NH_3]^2}{(k_1^o[N_2] + k_{-2}^o[NH_3]^2)^m (k_2^o[H_2]^3 + k_{-1}^o)^{1-m}}\right) \qquad (2)$$

Because of the above inequalities, two of the terms in the denominator can be neglected and equation 2 simplifies to:

$$r = \vec{k}\,[N_2]([H_2]^3/[NH_3]^2)^m - \overleftarrow{k}\,([NH_3]^2/[H_2]^3)^{1-m} \qquad (3)$$

where $\vec{k} = L\tau k_1^o K_2^{o\,m}$ and $\overleftarrow{k} = L\tau k_{-1}^o K_2^{o(m-1)}$. Thus the rate constants are expressed in terms of rate parameters associated with the sites having the highest affinity.

In an effort to verify this model, this reaction was compared to that using deuterium to produce deuterated ammonia [16]:

(4)
$$N_2 + 3\,D_2 \rightleftharpoons 2\,ND_3$$

The ratio of the overall rate constants for reactions (3) and (4) in the forward direction far from equilibrium is:

$$\vec{k}_D/\vec{k}_H = (k_{1,D}^o/k_{1,H}^o)(K_{2,D}^o/K_{2,H}^o)^m \qquad (4)$$

Step 1 does not involve H or D, so it is independent of either and

$$k_{1,D}^o = k_{1,H}^o \tag{5}$$

Furthermore, note that the ratio $K_{2,D}^o / K_{2,H}^o$ is just the difference between the two equilibria, in other words, it represents the equilibrium for the isotopic exchange reaction:

(5)
$$N_2* + 3 D_2 \overset{K_{2,D}^o}{\rightleftharpoons} 2 ND_3 + *$$

(6)
$$* + 2 NH_3 \overset{1/K_{2,H}^o}{\rightleftharpoons} 3 H_2 + N_2*$$

_____

(7)
$$3 D_2 + 2 NH_3 \rightleftharpoons 3 H_2 + 2 ND_3$$

Reaction 7 involves only gas-phase compounds and the equilibrium constants can be calculated from free energies listed in thermodynamic tables, i.e.,

$$K_{2,D}^o / K_{2,H}^o = [H_2]^3 [ND_3]^2 / [D_2]^3 [NH_3]^2 \tag{6}$$

Finally, examination of equations 4 and 5 shows

$$\vec{k}_D / \vec{k}_H = (K_{2,D}^o / K_{2,H}^o)^m \tag{7}$$

Using an iron catalyst which had an experimental value of $m = 1/2$, Shapatina et al. measured these forward rate constants at three temperatures and compared them to those calculated using equation 7, and the results are shown below in Table 1 [17]. The agreement is remarkably good, especially when it is recognized that no adjustable parameters were involved.

_____

Before the topic of nonuniform surfaces is concluded, it is interesting to compare the rate equation obtained by Temkin and Pyzhev for ammonia synthesis on iron (and discussed in Illustration 8.1) to one associated with a uniform surface using the same reaction model. Comparing only the forward rate in either sequence, one would have

TABLE 8.1. Comparison of experimental and theoretical rate parameters in the Temkin-Pyzhev rate expression for NH_3 synthesis (Equation 2 with $m = 1/2$). (from ref. 3)

Temperature (K)	$\vec{k}_D / \vec{k}_H$ (experimental)	$\vec{k}_D / \vec{k}_H = \left(K_{2,D}^o / K_{2,H}^o \right)^{1/2}$ (calculated)
673	3.13	2.87
723	2.89	2.60
748	2.61	2.52

$$(8.5) \qquad N_2 + * \xrightarrow{\quad k_A \quad} N_2*$$

$$(8.6) \qquad N_2* + 3 H_2 \xleftrightharpoons{\quad K \quad} 2 NH_3 + *$$

For a H-W-type reaction sequence on a uniform surface

$$r = k[N_2][*] \qquad (8.68)$$

If N_2* is again assumed to be the MARI, which is admittedly a very precarious assumption, but consistent with that in Illustration 8.1, then the site balance is $L = [*] + [N_2*]$ and

$$r = Lk[N_2]/(1 + [NH_3]^2/K[H_2]^3) \qquad (8.69)$$

If this expression is approximated by a power series, then

$$r = k'[N_2]([H_2]^3/[NH_3]^2)^n \qquad (8.70)$$

where $0 < n < 1$. Consequently, the final mathematical form is identical to the first term in equation 3 in Illustration 8.1.

As a consequence of the ambiguity of a two-step sequence [10], other two-step models with perhaps more satisfying assumptions can also give the same final rate expression. For example, for the above irreversible reaction assume that *dissociative* N_2 adsorption is the RDS and adsorbed N *atoms* are the MARI, i.e.,

$$(8.7) \qquad N_2 + 2 * \xrightarrow{\quad k_1 \quad} 2 N*$$

$$(8.8) \qquad 2 [N* + 3/2 H_2 \xleftrightharpoons{\quad K_2 \quad} NH_3 + *]$$

$$\overline{}$$

$$(8.9) \qquad N_2 + 3 H_2 \implies 2 NH_3$$

For a uniform surface the forward rate is

$$r = k_1[N_2][*]^2 \qquad (8.71)$$

and the site balance is

$$L = [*] + [N*] = [*] + [*][NH_3]/K_2[H_2]^{3/2} \qquad (8.72)$$

Thus the forward rate equation is

$$r = Lk'[N_2]/(1 + K'[NH_3]/[H_2]^{3/2})^2 \qquad (8.73)$$

which, if again approximated by a power series (or if the coverage of N atoms is very high), gives a final rate expression identical to equation 8.70.

References

1. F. H. Constable, *Proc. Royal Soc. London A* 108 (1925) 355.
2. M. Boudart, "Kinetics of Chemical Processes", Prentice Hall, Englewood Cliffs, NJ, 1968.
3. M. Boudart and G. Djéga-Mariadassou, "Kinetics of Heterogeneous Catalytic Reactions", Princeton Press, Princeton, NJ, 1985.
4. M. G. Evans and M. Polanyi, *Trans. Faraday Soc.* 34 (1938) 11.
5. N. N. Semenov, "Some Problems in Chemical Kinetics and Reactivity", Pergamon, London, 1958.
6. (a) J. N. Brønsted and K. Pederson, *Z. Physik Chem.* 108 (1923) 185.
   (b) J. N. Brønsted and V. K. LaMer, *J. Am. Chem. Soc.* 46 (1924) 555.
7. R. P. Bell, "The Proton in Chemistry", Methuen & Co., Ltd., London, 1959.
8. L. P. Hammett, "Physical Organic Chemistry", McGraw-Hill, NY, 1940.
9. J. Hine, "Physical Organic Chemistry", McGraw-Hill, NY, 1963.
10. M. Boudart, *AIChE J.* 18 (1972) 465.
11. M. Boudart in "Physical Chemistry", H. Eyring, D. Henderson and W. Jost, Eds., Chapter 7, Academic Press, NY, 1975.
12. M. I. Temkin, *Zhur. Fiz. Khim.* 31 (1957) 1.
13. M. I. Temkin, *Dok. Akad. Nauk. SSSR* 161 (1965) 160.
14. M. I. Temkin, *Adv. Catal.* 28 (1979) 173.
15. M. I. Temkin and V. Pyzhev, *Acta Physicochem. URSS* 12 (1940) 217.
16. A. Ozaki, H. Taylor and M. Boudart, *Proc. Royal Soc. (London)* A258 (1960) 47.
17. E. I. Shapatina, V. L. Kuchaev and M. I. Temkin, *Kinet. i Katal.* 12 (1971) 1476.

Problem 8.1

What is the optimum fractional coverage of a nonuniform catalyst surface if the transfer coefficient α is 2/3 rather than 1/2; in other words, if equation 8.57 utilizes $\alpha = 2/3$?

Problem 8.2

The Temkin rate equation for NH_3 synthesis is based on the 2-step sequence provided in Illustration 8.1 and it is given by equation 2 in that Illustration. With the information given there, i.e., $m = \alpha = 1/2$, verify that equation 3 can be derived and that the listed values of $\overrightarrow{k}$ and $\overleftarrow{k}$ are obtained.

9
Kinetics of Enzyme-Catalyzed Reactions

Chemical reactions between biochemical compounds are enhanced by biological catalysts called enzymes, which consist mostly or entirely of globular proteins. In many cases a cofactor is needed to combine with an otherwise inactive protein to produce the catalytically active enzyme complex. The two distinct varieties of cofactors are coenzymes, which are complex organic molecules, and metal ions. Enzymes catalyze six major classes of reactions: 1) Oxidoreductases (oxidation-reduction reactions), 2) Transferases (transfer of functional groups), 3) Hydrolases (hydrolysis reactions), 4) Lyases (addition to double bonds, 5) Isomerases (isomerization reactions) and 6) Ligases (formation of bonds with ATP (adenosine triphosphate) cleavage) [1].

It is frequently stated that enzymes are more active than synthesized inorganic catalysts and, consequently, have much higher TOFs. This may be true in the low-temperature region where these enzymes typically operate, but their activity does not increase continuously with temperature, thus the range of operating temperature is quite limited. Solid inorganic catalysts can give TOFs at higher temperatures that are comparable to, or higher than, those with enzymes [2]. Regardless of the greater complexity that can occur with enzymes, the definitions and concepts employed previously for heterogeneous catalysts are still applicable in these systems. Thus the use of TOFs, active site balances, quasi-equilibrated steps, and the steady-state approximation (SSA) is valid for these biological systems, and these concepts are regularly employed. An excellent discussion of the topic of enzyme kinetics is provided by Bailey and Ollis [3].

9.1   Single-Substrate Reactions

Because of the high activity of enzymes, the SSA might be expected to be especially useful and, indeed, it provides the most general form of the rate expression [3]. A simple sequence of elementary steps that frequently describes an enzyme-catalyzed biological process converting reactant A (referred to as a substrate in the biochemistry literature) to product P is the following, where E is the enzyme:

$$(9.1) \qquad\qquad A + E \;\underset{k_{-1}}{\overset{k_1}{\rightleftharpoons}}\; A - E$$

$$(9.2) \qquad\qquad A - E \;\xrightarrow{k_2}\; P + E$$

$$(9.3) \qquad\qquad A \;\Longrightarrow\; P$$

The rate of the overall process given by reaction 9.3 is:

$$r = -d[A]/dt = d[P]/dt = k_2[A - E] \tag{9.1}$$

If the rate of step 9.2 is high enough compared to the forward and reverse rates of step 9.1, then the latter step cannot be assumed to be quasi-equilibrated and the SSA must be employed to eliminate the unknown concentration of the active complex [A−E], i.e.,

$$d[A - E]/dt = k_1[A][E] - k_{-1}[A - E] - k_2[A - E] = 0 \tag{9.2}$$

thus

$$[A - E] = k_1[A][E]/(k_{-1} + k_2). \tag{9.3}$$

The active site balance on the enzyme, where L_e is the total enzyme concentration, is:

$$L_e = [E] + [A - E] \tag{9.4}$$

Solving these two equations simultaneously gives:

$$[A - E] = \frac{L_e k_1[A]}{k_{-1} + k_2 + k_1[A]} \tag{9.5}$$

and substituting this into equation 9.1 provides the final rate law:

$$r = \frac{L_e k_1 k_2[A]}{k_{-1} + k_2 + k_1[A]} = \frac{L_e k_2[A]}{\dfrac{k_{-1} + k_2}{k_1} + [A]} \tag{9.6}$$

This derivation was first proposed by Briggs and Haldane in 1925 [4]. The latter representation of this rate law has the same mathematical form as that found experimentally by Henri in 1902 [5] and by Michaelis and Menten in 1913 [6], which was originally expressed as:

$$r = \frac{r_{max}[A]}{K_m + [A]} \tag{9.7}$$

This has become known as the Michaelis-Menten equation, with r_{max} representing the maximum or limiting rate and K_m being called the Michaelis constant. Equation 9.7 has been frequently expressed as:

$$v = \frac{VA}{K+A} = \frac{v_{max}S}{K_m+S} \qquad (9.8)$$

where A or S represents the concentration of reactant A or S, respectively, and its form with the rate expressed as v, shown in Figure 9.1, is analogous to that in Figure 7.1(a). This equation can be readily linearized to give:

$$\frac{1}{v} = \left(\frac{K}{V}\right)\left(\frac{1}{A}\right) + \frac{1}{V}, \qquad (9.9)$$

which is commonly known as a Lineweaver-Burke plot [7]. If the data in Figure 9.1 are converted to such a reciprocal plot, the result in Figure 9.2 is obtained [8].

Some important comments can be made about equation 9.7. First, $r_{max} = L_e k_2$ and it is achieved when $[A] >> K_m$; furthermore, the rate is one-half its maximum value when $[A] = K_m$. Second, as might be expected based on a 2-step sequence [9], other assumptions might produce an identical form of the rate expression and, indeed, the first derivation by the above authors invoked quasi-equilibrium for step 9.1, which gives equation 9.7, but K_m now equals k_{-1}/k_1, rather than $(k_{-1}+k_2)/k_1$, and it has the meaning of a dissociation constant for the active complex [5,6]. In this case, both the catalytic cycle and the final rate equation are analogous to a unimolecular L-H reaction. (Note that dividing both numerator and denominator in

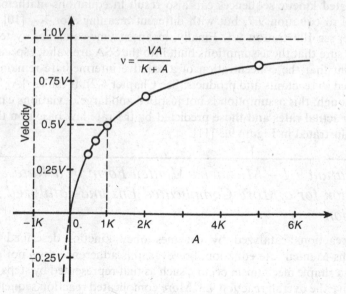

FIGURE 9.1. Typical form of Michaelis-Menton kinetic results, as represented by Eq. 9.7 or Eq. 9.8. (Reproduced from ref. 8, copyright © 1972, with permission of the McGraw-Hill Companies)

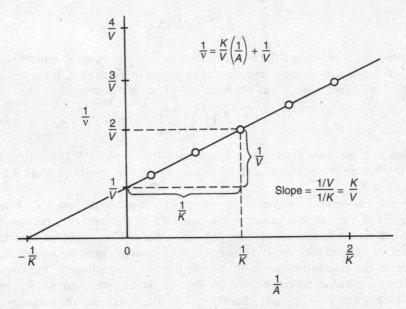

FIGURE 9.2. Lineweaver-Burke plot corresponding to Figure 9.1. (Reprinted from ref. 8, copyright © 1972, with permission of the McGraw-Hill Companies)

equation 9.7 by K_m gives the L-H form of the rate law.) Furthermore, more complicated kinetic sequences can also result in equations mathematically identical to equation 9.7, but with different meanings for K_m [10]. As an example, see Illustration 9.1. Finally, in an enzyme-catalyzed system, one must be sure that the assumptions built into the SSA are valid, especially the constraint that the concentration of a reactive intermediate remains small compared to reactants and products (See Chapter 6.2). If the $L_e/[A]_o$ ratio is large enough, this assumption is not justified and large deviations can occur between actual rates and those predicted by the rate law based on the SSA [3], as illustrated in Figure 9.3 [11].

Illustration 9.1 – Michaelis-Menten Form of a Rate Equation for a More Complicated Enzyme-catalyzed Reaction

Many reactions catalyzed by enzymes obey kinetics described by the Michaelis-Menten rate equation; however, this adherence does not guarantee that a simple mechanism occurs, such as that represented by steps 9.1 and 9.2 to give the overall reaction 9.3. More complicated reaction sequences can result in exactly the same kinetic behavior. For example, there is considerable evidence that the following mechanism describes a number of enzyme

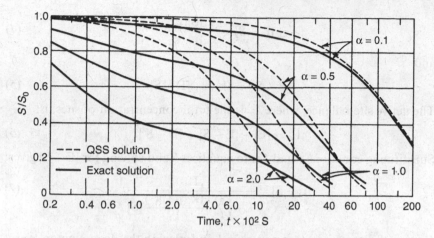

FIGURE 9.3. Computed time course of batch hydrolysis of acetyl 1-phenylalanine ether by chymotrypsin. Considerable discrepancies arise between the exact solution and the quasi-steady-state (i.e., the SSA) solution when $\alpha = e_o/S_o = L_e/[A]_o$ is not sufficiently small. (From ref. 11, copyright © 1973 AIChE, reproduced with permission of the American Institute of Chemical Engineers)

systems [10], where S represents the substrate, E is the enzyme, and Y and Z are the products:

(1)
$$S + E \underset{k_{-1}}{\overset{k_1}{\rightleftharpoons}} E - S$$

(2)
$$E - S \xrightarrow{k_2} E - S' + Y$$

(3)
$$E - S' \xrightarrow{k_3} Z + E$$

---

(4)
$$S \Longrightarrow Y + Z$$

The rate of this reaction is:

$$r = -d[S]/dt = d[Y]/dt = d[Z]/dt = k_2[E - S] = k_3[E - S'] \qquad (1)$$

Here there are two reactive intermediates, E-S and E-S′, and application of the SSA to each gives:

$$d[E - S]/dt = k_1[S][E] - (k_{-1} + k_2)[E - S] = 0 \qquad (2)$$

and

$$d[E - S']/dt = k_2[E - S] - k_3[E - S'] = 0 \qquad (3)$$

These lead to the relationships

$$[E - S] = \frac{k_1[S][E]}{k_{-1} + k_2} \tag{4}$$

and

$$[E - S'] = (k_2/k_3)[E - S] \tag{5}$$

The active site balance, where L_e is the total concentration of sites, is:

$$L_e = [E] + [E - S] + [E - S'] \tag{6}$$

Substituting equations 4 and 5 into equation 6 and solving for $[E - S]$ gives:

$$[E - S] = \frac{L_e}{1 + \dfrac{k_2}{k_3} + \dfrac{(k_{-1} + k_2)}{k_1}[S]} \tag{7}$$

Putting this term into rate equation 1 and placing the denominator over a common denominator results in:

$$r = L_e k_2 \bigg/ \left(\frac{k_1[S] + (k_1 k_2/k_3)[S] + (k_{-1} + k_2)}{k_1[S]} \right) \tag{8}$$

which can be rearranged to give:

$$r = \frac{L_e k_1 k_2 [S]}{k_{-1} + k_2 + \left(\dfrac{k_1 k_2 + k_1 k_3}{k_3} \right)[S]} \tag{9}$$

Finally, dividing both numerator and denominator by the factor multiplying [S] in the denominator produces the final Michaelis-Menten form of the rate equation, i.e.,

$$r = \frac{\left(\dfrac{L_e k_2 k_3}{k_2 + k_3} \right)[S]}{\left(\dfrac{k_{-1} k_3 + k_2 k_3}{k_1 k_2 + k_1 k_3} \right) + [S]}. \tag{10}$$

The interpretation of the two constants in equation 10 corresponding to r_{max} and K_m in equation 9.7 is clearly more complicated than that for the simple mechanism discussed previously.

9.2   Dual-Substrate Reactions

Additional analogies occur between reactions catalyzed by enzymes and by surfaces when reactions between two substrates are considered. A large majority of reactions catalyzed by enzymes involve at least two substrates; however, one is frequently water, whose concentration is typically much larger than that of the other substrates and therefore remains essentially constant, thus simplifying the kinetics. For example, a dual-substrate reaction with water as one of the substrates could be treated as discussed in the previous section.

Many dual-substrate reactions can be represented by a sequence of elementary steps involving a ternary complex comprised of the enzyme and the two reactants, and application of the SSA to these systems can be quite complicated [10]. If a RDS is assumed to exist, then the derivation of a rate expression can be markedly simplified, as shown next. Assume that substrates A and B interact with an enzyme E to form a product P according to the following series of elementary steps, where the first four steps are quasi-equilibrated and the last step is the RDS:

$$(9.4) \qquad A + E \underset{}{\overset{K_A}{\rightleftharpoons}} E - A \qquad (QE)$$

$$(9.5) \qquad B + E \underset{}{\overset{K_B}{\rightleftharpoons}} E - B \qquad (QE)$$

$$(9.6) \qquad E{-}A + B \underset{}{\overset{K_{AB}}{\rightleftharpoons}} E - AB \qquad (QE)$$

$$(9.7) \qquad E{-}B + A \underset{}{\overset{K_{BA}}{\rightleftharpoons}} E - AB \qquad (QE)$$

$$(9.8) \qquad 2\,[E{-}AB \overset{k}{\underset{}{\longrightarrow}} P + E] \qquad (RDS)$$

$$(9.9) \qquad A + B \Longrightarrow P$$

The rate can be defined as:

$$r = d[P]/dt = k[E - AB] \qquad (9.10)$$

The site balance for the total enzyme concentration, L_e, is:

$$L_e = [E] + [E - A] + [E - B] + [E - AB] \qquad (9.11)$$

From the four quasi-equilibrated steps, one has, respectively:

$$[E - A] = K_A[A][E] \qquad (9.12)$$

$$[E - B] = K_B[B][E] \qquad (9.13)$$

$$[E - AB] = K_{AB}[E - A][B] \qquad (9.14)$$

$$[E - AB] = K_{BA}[E - B][A] \qquad (9.15)$$

Substitution of equations 9.12–9.15 into equation 9.11 and solving for [E] gives:

$$[E] = \frac{L_e}{1 + K_A[A] + K_B[B] + K_A K_{AB}[A][B]} \tag{9.16}$$

Substituting this into equation 9.12, then this equation into equation 9.14, and finally this last equation into equation 9.10 gives a final rate expression, after noting that $K_A K_{AB} = K_B K_{BA}$ because equations 9.14 and 9.15 are equal, of:

$$r = \frac{L_e k[A][B]}{K_A^{-1} K_{AB}^{-1} + K_{AB}^{-1}[A] + K_{BA}^{-1}[B] + [A][B]} \tag{9.17}$$

or, alternatively,

$$r = \frac{L_e k}{1 + K_{BA}^{-1}/[A] + K_{AB}^{-1}/[B] + K_A^{-1} K_{AB}^{-1}/[A][B]} \tag{9.18}$$

Note that the reciprocal of each K_i value can be viewed as a dissociation equilibrium constant for a particular complex.

If equation 9.17 or 9.18 is rearranged in the form of equation 9.7, i.e.,

$$r = \frac{r_{max}^* [A]}{K_m^* + [A]} \tag{9.19}$$

then

$$r_{max}^* = \frac{L_e k[B]}{K_{AB}^{-1} + [B]} \tag{9.20}$$

and

$$K_m^* = \frac{K_{BA}^{-1}[B] + K_A^{-1} K_{AB}^{-1}}{K_{AB}^{-1} + [B]} \tag{9.21}$$

It can be seen that if the concentration of one substrate is much larger than the other and remains essentially constant, then equation 9.19 will behave as a Michaelis-Menten rate law. The participation of a cofactor in a single-substrate enzymatic reaction (or a dual-substrate enzymatic reaction with $[B] \gg K_{AB}^{-1}$) can be modeled via the sequence given in steps 9.4-9.9. If the substrate concentration is considered to be essentially constant, then equation 9.19 exhibits a Michaelis-Menten dependence on cofactor concentration.

In concluding this short chapter on simple enzyme kinetics, several other aspects should be mentioned. First, the influence of pH as well as other activity modulators, particularly inhibitors and poisons, can be quantitatively accounted for using the approaches introduced here that can produce Michaelis-Menten-type rate expressions. Second, the apparent rate constant from many of these rate expressions obeys an Arrhenius form over a limited temperature range, but if the temperature becomes too high (ca. 325 K), the enzymes denature (fall apart). Finally, there has been, and continues to be, much interest in enzymes immobilized on surfaces, especially those of high-surface-area solids, and with these systems the concern of mass transfer limitations must again be recognized. Additional unique complications such as denaturation of the supported enzymes due to shear forces and enzyme loss caused by abrasion among particles must also be considered. If interested in these and other topics related to enzymatic kinetics, the reader is referred to the book by Bailey and Ollis [3].

References

1. A. L. Lehninger, "Biochemistry", 2nd Ed., Worth, NY, 1975.
2. R. W. Maatman, *Catal. Rev.* 8 (1973) 1.
3. J. E. Bailey and D. F. Ollis, "Biochemical Engineering Fundamentals", McGraw-Hill, NY, 1977.
4. G. E. Briggs and J. B. S. Haldane, *Biochem. J.* 19 (1925) 338.
5. V. Henri, "Lois Générales de l'Action des Diastases", Hermann, Paris, 1903.
6. L. Michaelis and M. L. Menten, *Biochem Z.* 49 (1913) 333.
7. H. Lineweaver and D. Burke, *J. Am. Chem. Soc.* 58 (1934) 658.
8. K. M. Plowman, "Enzyme Kinetics", McGraw-Hill, NY, 1972.
9. M. Boudart, *AIChE J.* 18 (1972) 465.
10. K. J. Laidler, "Chemical Kinetics", 3rd Ed., Harper and Row, NY, 1987.
11. H. C. Lim, *AIChE J.* 19 (1973) 659.

Problem 9.1

Derive the rate expression for an enzyme-catalyzed unimolecular (single substrate) reaction, such as that shown in steps 9.1 and 9.2, assuming that the decomposition of the reactive intermediate to give the product is reversible, rather than irreversible as indicated in step 9.2. Can the initial rate in the forward direction and the initial rate in the reverse direction be expressed in the form of a Michaelis-Menten rate equation? If so, how? If not, why?

Problem 9.2 (from ref. 3)

Initial rates of an enzyme-catalyzed reaction for various substrate (reactant) concentrations are listed in the table below. Evaluate r_{max} and K_m by a Lineweaver-Burke plot.

[A] (mole L^{-1})	r(mole min^{-1} L^{-1}) $\times 10^6$
4.1×10^{-3}	177
9.5×10^{-4}	173
5.2×10^{-4}	125
1.03×10^{-4}	106
4.9×10^{-5}	80
1.06×10^{-5}	67
5.1×10^{-6}	43

Problem 9.3 (from ref. 3)

Derive an expression for the reaction rate, r, in terms of S, E and the constants shown for the following reaction sequence, which includes substrate inhibition:

(1)
$$E + S \underset{}{\overset{K_s}{\rightleftharpoons}} ES$$

(2)
$$ES + S \underset{}{\overset{K'_s}{\rightleftharpoons}} ESS$$

(3)
$$ES \xrightarrow{\ k\ } E + P$$
$$\overline{\qquad\qquad\qquad\qquad}$$
$$S \Longrightarrow P$$

State all assumptions. E represents the enzyme.

Problem 9.4 (from ref. 3)

Multiple complexes can be involved in some enzyme-catalyzed reactions. For the reaction sequence shown below, develop suitable rate expressions using: (a) the Michaelis equilibrium approach and (b) the steady-state approximation for the complexes.

(1)
$$S + E \underset{k_{-1}}{\overset{k_1}{\rightleftharpoons}} (ES)_1$$

(2)
$$(ES)_1 \underset{k_{-2}}{\overset{k_2}{\rightleftharpoons}} (ES)_2$$

(3)
$$(ES)_2 \xrightarrow{\ k_3\ } P + E$$
$$\overline{\qquad\qquad\qquad\qquad}$$
$$S \Longrightarrow P$$

Problem 9.5 (from ref. 3)

The catalytically active form of an enzyme can depend on its state of ionization; consequently, the pH ($pH = -\log[H^+]$ of the reacting medium can have a significant effect on the rate of reaction. Assuming that the active form of an enzyme is that after the loss of a single proton, determine the influence of the proton concentration on the maximum reaction rate. The equations governing the two inactive forms, E and E^{-2}, due to the protonation and deprotonation of the active species, E^-, are, respectively:

(1)
$$E \underset{}{\overset{K_1}{\rightleftharpoons}} E^- + H^+$$

(2)
$$E^- \underset{}{\overset{K_2}{\rightleftharpoons}} E^{-2} + H^+$$

Such ionization reactions are very rapid compared with most reaction rates in solution, thus quasi-equilibrium can be assumed.

Subject Index